THE BIOLOGY
OF NECTARIES

THE BIOLOGY
OF NECTARIES

**BARBARA BENTLEY AND
THOMAS ELIAS, EDITORS**

COLUMBIA UNIVERSITY PRESS
NEW YORK 1983

Library of Congress Cataloging in Publication Data
Main entry under title:

The Biology of nectaries.

  Includes bibliographies and index.
  1. Nectaries—Addresses, essays, lectures.
I. Bentley, Barbara.   II. Elias, Thomas S.
QK657.B56   582′.014   82-4200
ISBN 0-231-04446-1   AACR2

Columbia University Press
New York   Guildford, Surrey

CONTENTS

PREFACE

Although the occurrence of nectaries on plants has been known since the time of Vergil, their physiological, ecological, and evolutionary roles remain an area of active research even today. In fact, legitimate research (as opposed to armchair philosophizing) on nectaries is a relatively recent phenomenon. For example, as recently as 1970, the ecological role of extrafloral nectaries was considered a mystery by some physiologists and even the correct identification of nectiferous tissues escaped some morphologists in their studies of major taxa. On the other hand, recent advances in instrumentation, especially the electron microscope, and various analytical techniques have allowed tremendous progress in these very areas. Hence the genesis of this volume.

The contributors we selected provide a representation from as wide a variety of nectary biologists as possible—ranging from the ultrastructure of nectaries through to the role of insect-plant interactions in agricultural systems. The papers also represent the multidisciplinary nature of studies on nectaries. An ecologist's field study can reveal the locations of nectiferous tissues for the morphologist to describe. And the physiologist's data can provide the mechanism for an evolutionist's hypothesis. In addition, we have tried to give fair representation to both temperate and tropical systems. The inclusion of the tropics represents not only the biological reality of the importance of nectaries in lower latitudes, but it will, we hope, encourage future research efforts in these areas where destruction of natural ecosystems is rampant and yet human needs are increasingly pressing.

From its conception at the AIBS symposium in 1977, through to the final illustration in this volume, we have been encouraged by the fascination for the study of nectaries, and we look forward to continuing research with equal fascination.

August 1981

B. L. Bentley
T. S. Elias

THE BIOLOGY
OF NECTARIES

1

THE ULTRASTRUCTURE OF FLORAL AND EXTRAFLORAL NECTARIES

LENORE T. DURKEE
GRINNELL COLLEGE

In the past few decades, considerable descriptive work has been done at the light microscope level on floral and extrafloral nectaries. The information derived from these studies has provided some partial answers to an array of problems ranging from the reasons for variation in nectar concentration (Frey-Wyssling and Agthe 1950) to questions of angiosperm phylogeny (Brown 1938). The studies have also revealed that great anatomical diversity exists among nectaries.

Since the early 1960s, the excellent work in light microscopy has been supplemented with the electron microscope to determine if there exists similar diversity in fine structure and to discover what ultrastructural changes might occur with the onset of secretion. Sometimes these investigations have included autoradiography or enzyme localization techniques and, in some especially well-documented species such as *Abutilon*, extensive physiological studies have been done on nectary tissue and nectar. Models for nectary secretion may be proposed using these various kinds of evidence.

Most nectaries consist of a mass of small, dense secretory or gland cells that may sometimes take the form of trichomes. The vascular tissue supplying the nectary does not come into contact with the secretory cells but is separated from them by one or more cells comprising the nonglandular or subglandular parenchyma (fig. 1.1). As one

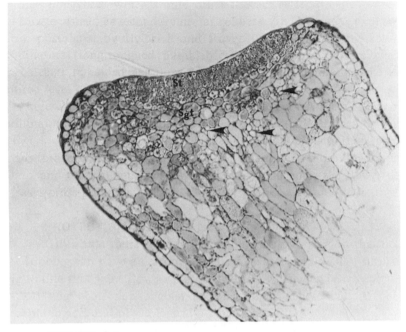

FIGURE 1.1
Passiflora coerulea L. Longitudinal section through a mature petiolar nectary showing the secretory (St), subglandular (Sgt), and vascular tissues (arrows). 200X.

St = secretory tissue   Sgt = subglandular tissue

surveys the literature, it becomes evident that much study has centered on the fine structure of the secretory or gland cell while very little attention has been given either to the nature of the subglandular tissue through which sugars must move to reach the secretory cells or to the vascular tissue which supplies these sugars. In addition, although many floral nectaries store starch, there have been few reports dealing with ultrastructural changes in this type of gland. Finally, as described elsewhere in this volume, it is now known that substances other than sugars appear to be widespread in the floral and extrafloral nectar of a variety of families. Therefore, information from microscope and physiological studies must be interpreted with this new knowledge in mind.

Fahn (1979a, 1979b) has extensively reviewed the ultrastructural studies of nectaries. This paper will briefly consider these findings, and will also examine what is known about subglandular tissue and the vascular supply of the nectaries. General models of nectary function will be discussed with possible new avenues for further research.

THE SECRETORY TISSUE

The tightly packed glandular cells, which are involved in the nectar secretion process, are characterized by a densely staining cytoplasm and relatively large nucleus (fig. 1.2). In this respect they

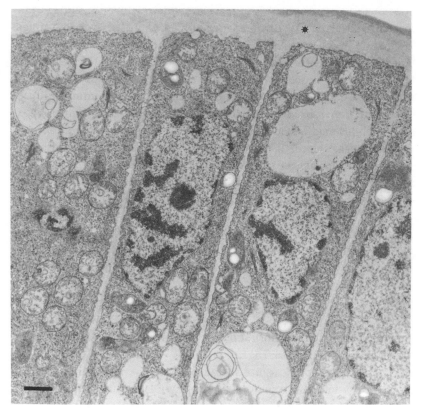

FIGURE 1.2
Passiflora coerulea. Secretory cells in an immature extrafloral nectary. The outer wall is indicated by *. 11,900X.

strongly resemble meristematic cells and it is the consistent high density and small size of these cells that enables the light microscopist to distinguish the secretory tissue from adjacent nonsecretory areas.

Electron microscopy reveals that the dense appearance results from an abundance of free ribosomes, membranes of both smooth and rough endoplasmic reticulum, and mitochondria. Other organelles are present but show considerable variation in abundance and appearance.

Endoplasmic Reticulum (ER) In almost all studies so far made, the ER is reported to be the most conspicuous feature in gland cells at the time of secretion (Wrischer 1962; Schnepf 1964b, 1964c; Eymé 1966; Fahn and Rachmilevitz 1970; Figier 1971; Findlay and Mercer 1971; Rachmilevitz and Fahn 1973; Tacina 1973; Benner and Schnepf 1975; Rachmilevitz and Fahn 1975; Durkee 1977; Baker et al. 1978; Fahn and Benouaiche 1979). The cisternae are sometimes described as dilated and either stacked in layers or scattered in the cytoplasm, often as vesicles. Both cisternae and vesicles are occasionally so closely associated with the plasmalemma that possible fusion of the components has been suggested (Fahn and Rachmilevitz 1970; Rachmilevitz and Fahn 1973, 1975). Some workers have observed continuities of ER cisternae with vacuoles (Eymé 1966; Zandonella 1970; Heinrich 1975b).

In contrast to these findings, however, Vasiliev (1971) reported no unusual amounts of endoplasmic reticulum in the floral nectaries of *Convolvulus* and *Heracleum*. Grout and Williams (1980) apparently did not observe extensive ER in the extrafloral nectaries of *Dioscorea* nor was it observed in the starch-storing floral nectaries of *Passiflora* spp (Durkee et al. 1981).

In the floral nectaries of *Diplotaxis* and *Helleborus*, Eymé and LeBlanc (1963) and Eymé (1966) observed structures composed of tightly convoluted tubules which they termed "cotte de maille." These are evidently modified cisternae of ER and Eymé suggested that they might represent deposits of excess membrane material synthesized by these cells. They seem to be characteristic of a variety of metabolically active cells (Bassot 1966; Wooding 1967; Camp and Whittingham 1972; Zachariah and Anderson 1973).

Vacuoles A reduced vacuome is said to be characteristic of active nectar-secreting cells (Rachmilevitz and Fahn 1975). For example, in *Gasteria* septal nectaries the vacuoles are inconspicuous during the secretory phase, but later they expand to form large central vacuoles when secretion has ceased (Schnepf 1973). However some workers have observed vacuolar expansion or a change from a few large vacuoles to numerous small ones as secretion commences (Wrischer 1962; Eymé 1963, 1966; Zandonella 1970; Benner and Schnepf 1975; Heinrich 1975a; Durkee 1977; Baker et al. 1978). In some cases the vacuoles are probably functioning as lysosomes (Zandonella 1970).

Mitochondria As might be expected, these organelles occur in great numbers and possess many well-developed cristae. They are usually distributed rather uniformly, but in those nectary cells characterized by wall protuberances (see below), the mitochondria may aggregate along the heavily invaginated plasmalemma (Schnepf 1964a). Large groups of mitochondria have also been reported in the vicinity of the nucleus in some floral nectaries (Eymé 1963). These assemblages were first detected by light microscopy and their significance is not known.

Dictyosomes Golgi bodies are usually conspicuous in the early growth period of the nectary before secretion has begun. They are particularly abundant in the pre-secretory stages of those nectaries in which wall labyrinths occur and are probably responsible for the formation of these structures. Dictyosome activity usually subsides with the onset of secretion in most species. However, there have been reports of sustained dictyosome vesicle production during the secretory phase (Eymé 1966; Figier 1971; Tacina 1973; Heinrich 1975a; Rachmilevitz and Fahn 1975; Fahn and Benouaiche 1979), and in the septal nectaries of some bromeliads unusually hypertrophied dictyosome vesicles have been reported during the time of maximum secretion (Benner and Schnepf 1975). These vesicles may fuse with the plasmalemma since they are often observed in proximity to this membrane.

Plastids   The secretory cell plastids of nectaries typically have poorly developed internal membrane systems. If starch is present prior to secretion it usually occurs only in small amounts and commonly disappears as secretion progresses. This situation exists in most floral and all extrafloral nectaries which have been examined at the ultrastructural level. In some species, however, the starch content of the floral nectaries and the numbers of mitochondria increase dramatically (fig. 1.3) at the approach of anthesis (Tacina 1973; Durkee et al. 1981). A rapid decline in starch coincident with the first appearance of nectar is observed in *Passiflora* (Durkee et al. 1981).

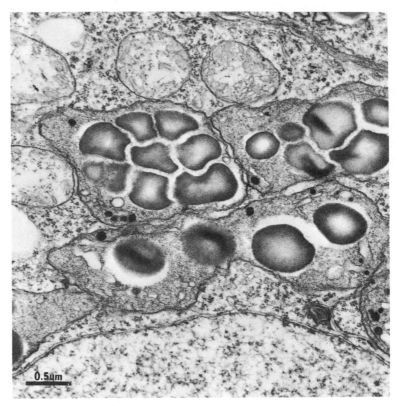

FIGURE 1.3
Passiflora trinifolia s.l. Floral nectary showing starch-storing secretory cell just prior to anthesis. 36,700X.

Large assemblages of plastids have been observed in proximity to the nucleus in the floral nectaries of some members of the Caryophyllaceae (Zandonella 1970). Their arrangement is similar to that described for mitochondria and their significance is unknown.

Plasmodesmata Secretory cells are well supplied with simple plasmodesmata in the radial and basal walls. Thus, adjacent secretory cells are connected to each other and to cells of the subglandular tissue. The plasmodesmata are usually more numerous and more complex in the walls between secretory and subglandular cells (Figier 1971; Durkee 1977). Some ER cisternae, often in the form of a complex network, are in proximity to the plasmodesmata and probably traverse the plasmodesmatal channel (Schnepf 1964a; Gunning and Hughes 1976).

Plasmalemma and Wall In some secreting nectaries, pronounced invaginations of the plasmalemma are observed (Eymé 1966; Fahn and Rachmilevitz 1970; Rachmilevitz and Fahn 1973; Benner and Schnepf 1975; Rachmilevitz and Fahn 1975; Durkee 1977). Sometimes the invaginations are associated with the plasmodesmata, as in the cross-walls of *Abutilon* nectary hairs (Findlay and Mercer 1971). In active nectaries as well as those which have ceased secretion, invaginations may be filled with small vesicles; that is the invaginations resemble lomasomes (Eymé 1966, 1967; Findlay and Mercer 1971). In other species, such modification of the plasmalemma is not observed. For example, in the cyathial nectaries of *Euphorbia pulcherrima* (Schnepf 1964c), the plasmalemma is smooth and lies close to the wall except in very young or very old nectaries. No invaginations were observed in the floral nectaries of various Caryophyllaceae (Zandonella 1970), *Acer platanoides* (Vasiliev 1969), or *Passiflora* (Durkee et al. 1981).

Some nectar-secreting cells contain conspicuous wall ingrowths which reach their maximum development with the onset of secretion (Wrischer 1962; Fahn and Rachmilevitz 1970; Figier 1971; Heinrich 1975a). These may disappear when secretion ceases as reported for *Gasteria* septal nectaries (Schnepf 1973).

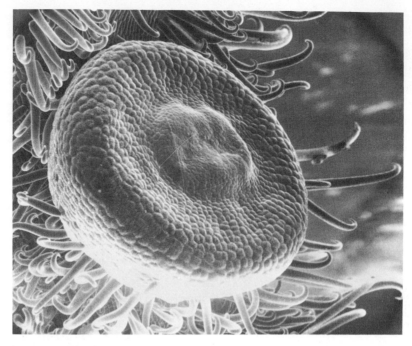

FIGURE 1.4
Passiflora warmingii Mast. Nearly mature extrafloral nectary. An accumulation of nectar can be seen as a bulge beneath the cuticle. 320X.

Fahn (1952) has suggested that nectar may diffuse through the thin secretory cell walls of some species. In other species, nectar is exuded through modified stomates (Rachmilevitz and Fahn 1975; Durkee et al. 1981). Commonly, nectar accumulates beneath the cuticle which eventually ruptures, releasing the sugary solution (fig. 1.4). In *Abutilon*, small pores serve as exit passages for accumulated nectar (Mercer and Rathgeber 1962). In this genus, additional heavy cutinization of the radial walls of the stalk cells is characteristic and apparently a critical factor in the method of nectar transport in these hairs. The stalk cells of *Lathraea* (Schnepf 1964a) and *Vicia faba* nectary trichomes (Wrischer 1962; Figier 1971), as well as the modified stalk of *Aphelandra* extrafloral nectaries (personal observation) have a similar wall structure (fig. 1.5).

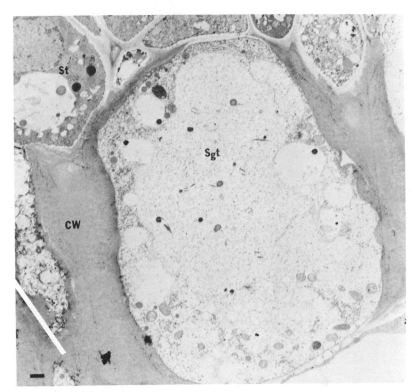

FIGURE 1.5
Aphelandra deppeana Schelcht. and Cham. Extrafloral nectary. The lateral walls of the subglandular cells (Sgt) are thick and heavily cutinized. These cells are less dense and contain fewer organelles than those of the secretory tissue (St). 5800X.

St = secretory tissue   Sgt = subglandular tissue   CW = cell wall

THE SUBGLANDULAR TISSUE

In most nectaries a subglandular parenchyma is present which consists of usually not more than three or four layers of isodiametric cells, separating the secretory layer from the phloem. The subglandular cells are typically larger than the secretory cells and are loosely packed. In *Passiflora* extrafloral nectaries, a single subglandular cell subtends three to four secretory cells in one plane of cut with the tissue three to four cells thick. In the floral nectary of this genus, the vascular tissue is embedded in the starch storing glandular area, but is separated from it by a sheath of parenchyma cells (Durkee 1977; Durkee et al. 1981).

Most evidence indicates that vascularized nectaries secrete sugars supplied from the nearby phloem (Agthe 1951). Even in *Gossypium*, where the subglandular area is so wide that the nectary lacks a special vascular supply, the secreted sugars may be transported from the phloem across this distance to the secretory cells (Wergin et al. 1975). It follows that the subglandular tissue, viewed as a pathway along which translocated material moves, may possess structural modifications for this function. Generally the ground plasma is reported to be less dense and neither endoplasmic reticulum nor the Golgi apparatus show the unusual degree of development and swelling found in the secretory tissue (Wrischer 1962; Fahn and Rachmilevitz 1970; Figier 1971; Findlay and Mercer 1971; Benner and Schnepf 1975; Durkee 1977). The plastids contain small amounts of starch and mitochondria may be abundant (Wergin et al. 1975; Baker et al. 1978) or sparse (Figier 1971; Durkee 1977).

The extent of vacuolation varies. In *Ricinus* extrafloral nectaries, large vacuoles are present in the pre-secretory stage. During secretion, many small vacuoles appear (Baker et al. 1978). In other cases, vacuoles are always well-developed (Figier 1971; Findlay and Mercer 1971; Wergin et al. 1975; Durkee 1977; Durkee et al. 1981). Reports of continuity between ER cisternae and vacuoles are difficult to establish.

Plasmodesmata are present in very large numbers (Figier 1971; Findlay and Mercer 1971). In *Passiflora*, they are often much-branched structures occupying enlarged pockets within the wall itself (Durkee 1977). In *Gossypium*, the plasmodesmata may be clustered in pit fields in concentrations as high as $40/um^2$ (Wergin et al. 1975). Cisternae of ER are often associated with the plasmodesmata, but intercellular continuities of ER through these channels are difficult to demonstrate.

THE VASCULAR TISSUE

As described earlier, in vascularized nectaries the vein endings generally terminate a few cells from the secretory tissue. They consist of phloem or, less commonly, both phloem and xylem. Few descriptions of the fine structure of nectary vascular tissue are available.

In *Vicia faba*, the stipular nectaries are supplied entirely by phloem (Figier 1971). The companion cells are Type A transfer cells (Gunning and Pate 1969) since they possess extensive wall ingrowths. In this and other cytological features, Figier has noted that the companion cells and secretory cells of *Vicia faba* are very similar. Characteristic plasmodesmata occur between companion cells and sieve elements, but are rarely seen between companion cells and adjacent parenchyma. His figure 3 indicates that the companion cells are considerably larger than the sieve elements.

The nectary phloem of *Abutilon* contains companion cells which evidently lack wall ingrowths. They are very dense cells with numerous organelles, especially dictyosomes, and are comparable in size to the sieve elements (Findley and Mercer 1971). Unfortunately, the distribution of plasmodesmata is not described. In *Gossypium*, the closest vascular tissue is that of the main vein (Wergin et al. 1975). Their figure 8 shows the size relationship between companion cell and sieve element. As would be expected in main path phloem, the companion cells are narrower and no unusual distribution of plasmodesmata was observed.

The studies of *Passiflora* floral and extrafloral nectaries (Durkee 1977; Durkee et al. 1981; Durkee, in press) indicate that the vascular tissue supplying the nectaries also terminates in phloem. In this area, the sieve elements and companion cells are conspicuously truncate and the sieve elements are typically narrower than the companion cells (fig. 1.6), an arrangement which is quite similar to that in leaf minor veins (Esau 1967). The typical plasmodesmata of the companion cell-sieve element complex are present but wall ingrowths are not seen. The companion cells contain p-protein and are very dense and unusually rich in organelles, particularly ribosomes, ER cisternae, and dictyosomes.

It is often difficult to distinguish between companion cells and adjacent phloem parenchyma. The latter cells, somewhat less dense, ultrastructurally resemble the companion cells. However, their nuclei are often lobed and closely associated with clusters of mitochondria. The presence of these aggregations confirm the earlier light micro-

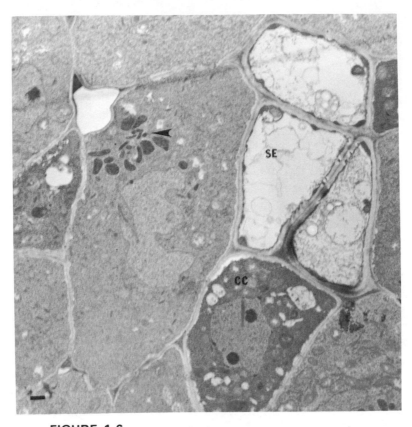

FIGURE 1.6

Passiflora biflora Lam. Extrafloral nectary. The termination of a vascular bundle is shown. The large phloem parenchyma/companion cell (see text) contains an assemblage of plastids (arrow). 5800X.

SE = sieve element   CC = companion cell

scope observations of the subglandular tissue of *Passiflora* (Cusset 1965). A common component of these cells is masses of fibrillar membrane-bound material which cytochemical tests have indicated is proteinaceous (fig. 1.7). In *Passiflora warmingii* extrafloral nectaries the phloem parenchyma is unusually hypertrophied and the protein material is conspicuous. In the pre-secretory and secretory phase, the material occupies much of the cell. In senescent nectaries it is reduced or absent. This membrane-bound protein is often found with p-pro-

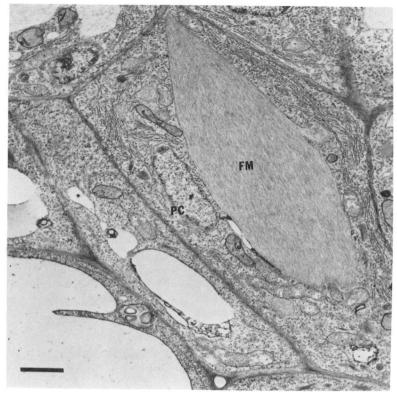

FIGURE 1.7

Passiflora warmingii. Floral nectary several days before anthesis. Proteinaceous fibrillar material within a membrane occupies a large volume of this phloem parenchyma cell. 17,500X.

FM = fibrillar material   PC = phloem parenchyma cell

tein in the same cell. Similar protein material has been observed in the phloem parenchyma of *Passiflora* floral nectaries (Durkee et al. 1981; Durkee, in press). In both cases, RER is abundant with cisternae in a parallel arrangement or sometimes branched. Although such protein accumulations have not been previously reported for nectaries, it is possible that they correspond to the finely dispersed material mentioned by Wrischer (1962) in the vacuoles of *Vicia faba* subglandular tissue and to the unidentified finely homogenous inclusions in the subglandular cell vacuoles of *Euphorbia* nectaries (Schnepf 1964c).

THE SECRETORY PROCESS

Interpretation of nectary ultrastructure cannot be divorced from physiological considerations. Nectar composition, its concentration, and the nature of the phloem sap from which it is derived are just a few of the factors that must logically fit into the ultrastructural framework to yield a reasonable model of the secretory process. There is an abundance of literature on nectary physiology and from this wealth of data some common features emerge.

1. Nectar composition. Percival (1961), in a study of nearly 900 species, observed that most floral nectar contains primarily glucose, fructose, and sucrose in proportions which vary interspecifically but show a high degree of constancy within a species. Although extrafloral nectars have not been studied to this extent, the kinds of sugars so far detected appear to be similar. In addition, oligosaccharides, proteins, amino acids, organic acids, and other substances may occur (Lüttge 1961; Zaurolov and Paulinova 1974; Baker and Baker 1975). There is mounting evidence that, like the sugars, the amino acid complement of floral and extrafloral nectars exhibits intraspecific constancy (Baker et al. 1978). Recent findings also indicate that the amino acid concentration in some nectars is considerably higher than in phloem sap (Pickett and Clark 1979; Inouye and Inouye 1980). The occurrence of these and other nitrogenous substances is not the result of bacterial action as has been suggested (Lüttge and Schnepf 1976) but rather nectary metabolism.

2. Nectar concentration. It has been shown that nectar concentration in a particular species is influenced by the nectary vascular supply; nectaries supplied by predominantly xylem, for example, produce a dilute nectar. Nectar concentrations therefore may vary from 8 percent sugar to 50 percent. (Frey-Wyssling and Agthe 1950). However, some nectars are reported to be as much as 75 percent sugar (Fahn 1979).

3. Phloem sap characteristics. The concentration of sugars in the phloem of over 500 tree species has been reported to range from less than 1 percent to 30 percent (Zimmerman and Ziegler 1975). Sucrose is the major sugar in all species so far studied. Since glucose and fructose are not present in phloem exudate (Zimmerman 1952; Kennecke et al. 1971) and if nectar is derived from phloem sap, the occurrence of these sugars indicates a degree of sucrose modification somewhere along the path from the sieve element to the exterior.

Comparisons of average sugar concentration of nectar and phloem sap have led to the conclusion that, in many cases, nectar is hypertonic to phloem sap. One of the few studies comparing nectar and phloem sap in the same species shows that the extrafloral nectar of *Ricinus* contains a much higher sugar concentration than the phloem (Baker et al. 1978). An extremely high concentration of carbohydrates has been also reported in the extrafloral nectar of *Opuntia* (Pickett and Clark 1979). Assuming that external conditions have not altered the nectar (by evaporation, for example), then some mechanism exists for concentrating the nectar—at the secretory cell membrane, in the secretory or subglandular cytoplasm, or at the level of the phloem.

Some resorption of substances by the nectary has been noted (Lüttge 1961; Baker et al. 1978). All these observations suggest that nectaries are often highly complex glands the functions of which we are only now beginning to appreciate.

With this information in mind, we are in a position to consider the two contrasting mechanisms which have been proposed to explain nectar secretion. In *granulocrine* secretion, a sugar solution is believed to be transported in vesicles to the plasmalemma of the secretory cell where fusion of the two components occurs, releasing nectar into the wall area. In *eccrine* secretion, individual sugar molecules are transported across the gland cell membrane, possibly by a carrier molecule. The evidence for both methods will be considered.

Granulocrine Secretion   The ER, its vesicles, and perhaps dictyosomes are believed by some to be involved in the secretory process because of their general abundance in gland cells (Eymé 1966, 1967; Fahn and Rachmilevitz 1970; Rachmilevitz and Fahn 1973; Heinrich 1975a, 1975b; Rachmilevitz and Fahn 1975). The frequently dilated appearance of the ER cisternae is taken to indicate the high osmotic potential within these structures resulting from an accumulation of sugars (Rachemilevitz and Fahn 1973) and their close association

with the plasmalemma suggests a possible fusion of these components. To determine if sugars were thus compartmentalized in the
ER, Fahn and Rachmilevitz (1975) fed tritiated sucrose to excised
flower buds and open flowers of *Lonicera*. Their autoradiographs
revealed label over some dilated ER cisternae and vesicles in actively
secreting cells whereas label was scarce in these cells during the pre-
secretory stage. Label was also observed over the plasmodesmata of
adjacent subglandular cells. Heinrich (1975a) fixed secreting *Aloe*
nectaries in vapors of glutaraldehyde and osmium after feeding the
tissue with tritiated glucose or ^{14}C-glucose. He was able to demonstrate
the presence of heavy silver deposits over the vacuoles and the nectar
cavity, but not in the ground plasm. The restriction of label to the
vacuoles, which were believed to be derived from ER or dictyosomes,
led to the conclusion that a granulocrine mechanism was operating.
However, Heinrich could not demonstrate with certainty that fusions
of vesicles with the plasmalemma were occurring. He also studied the
distribution of various phosphatases in the secreting *Aloe* nectary
(Heinrich 1975b). Although β-glycerophosphatase was present
throughout the cells, he found high activity of ATPase and glucose-6-
phosphatase which was apparently confined to the endoplasmic reticulum, dictyosome vesicles, and vacuoles. Those results, together
with the general lack of activity of these enzymes in the plasmalemma,
lent support to his belief that the ER and dictyosomes are important
in sugar transport via a granulocrine system. He suggested that sugars
are dephosphorylated in or at the surface of the cisternae and then
transported in vesicles to the plasmalemma where they are released.

The assumption that the endoplasmic reticulum can serve as a ve-
hicle for sugar transport in plant cells receives some support from
animal tissue study. Glycogen particles in liver cells are enveloped
by cisternae of smooth endoplasmic reticulum which become dilated
as the glycogen is converted to glucose. It is suggested that the ER
serves as the path for glucose movement in the cell; there appears to
be a fusion of these cisternae with the cell membrane (Porter 1963).
Glucose-6-phosphatase has also been detected in the cisternal membranes (Luck 1961).

In plant cells dictyosomes are primarily involved in the secretions
of polysaccharides (Morré et al. 1967; Schnepf 1968, 1969), and they

are ordinarily not active in nectary cells. Their enhanced activity in some nectaries, however, has raised the possibility that they may function in sugar secretion (Eymé 1966; Tacina 1973; Benner and Schnepf 1975; Heinrich 1975). It has been hypothesized that in the early stages of nectary ontogeny, especially in those in which wall ingrowths develop, dictyosomes provide polysaccharides for wall synthesis. With the onset of secretion, there may be a shift from insoluble to soluble carbohydrates in the dictyosome vesicles which then discharge their contents into the apoplast (Eymé 1966; Rachmilevitz and Fahn 1975). It should be noted here that some nectars, notably those of certain bromeliads, may contain polysaccharides (Budnowski 1922). Recognizing that the increased dictyosome activity in the bromeliad nectaries they were studying might reflect the formation of polysaccharides, Benner and Schnepf (1975) specifically tested for the presence of these compounds and found none. Thus they were led to conclude that the dictyosomes were secreting sugars.

In granulocrine secretion, the sugar concentrating step presumably occurs within the cisternae of endoplasmic reticulum or dictyosome vesicles or in the phloem although this problem has not been specifically addressed.

Eccrine Secretion The difficulty of establishing with certainty that secretory vesicles of ER origin fuse with the plasmalemma has been a factor leading many workers to discount this method of nectar secretion. For example, Schnepf (1964c) has reported the presence of well-developed, inflated, irregular profiles of rough endoplasmic reticulum and dictyosomes in the secreting nectary of *Euphorbia pulcherrima*. However, the absence of fusion of these structures with the plasmalemma and the fact that cooling the nectary, which halted secretion, did not alter the cytology of the cell, favored the conclusion that eccrine processes were operating in this tissue. In the starch-storing floral nectaries of *Passiflora*, neither ER nor dictyosomes are conspicuous features of the secretory or subglandular tissue. Mitochondria, however, become very abundant just prior to anthesis, and an eccrine secretion system is believed to be operating (Durkee et al. 1981).

Evidence that nectar secretion is slowed or halted by anaerobic conditions, cooling, and respiratory inhibitors and that nectar is often hypertonic to phloem sap constitutes a strong argument for the participation of membrane transport systems (Huber 1956; Ziegler 1956; Schnepf 1965c; Findlay et al. 1971). Thus, considerable emphasis has been placed on the presence of acid phosphatase in nectary cells, especially in the plasmalemma, as an indication of ATPase-mediated transport of sugars (Ziegler 1956; Vis 1958; Frey-Wyssling and Hauserman 1960; Figier 1968, 1971). As will be seen, there are some problems with this approach. It is reassuring, however, that more specific types of information are being obtained relating to sugar transport in higher plants and there is now a substantial body of evidence for a sucrose-specific transport system involving proton flux in various kinds of plant tissue (Komor 1977; Malek and Baker 1977, 1978; Racusen and Galston 1977; Hutching 1978; Giaguinta 1979). The existence of such sucrose pumps can not only explain the mechanism of phloem loading, but suggests that similar pumps could occur in nectary tissue plasma membranes to create the concentrated sugar solution. The wall proliferations which are characteristic of some nectaries (Schnepf 1964b; Figier 1971) are believed to enhance transport by increasing the surface area of the plasmalemma; these cells act like transfer cells in the intensive short distance movement of sugars. The ER is presumably the site for manufacture of those enzymes necessary for the transport process.

It has often been pointed out that the preponderance of sugars and the paucity of nonsugar components in nectar argues for eccrine processes with the plasmalemma exercising considerable control over the types of components secreted (Lüttge and Schnepf 1976). The selective retention of nitrogenous materials and specific ions has been correlated with increasing complexity of the nectary (Lüttge 1961). The accumulating evidence that often large amounts of nonsugar components such as amino acids may be secreted by complex glands necessitates a review of this proposal, but it is quite possible that these substances also may be exported by specific membrane transport systems as has been demonstrated for a variety of fungi and for some higher plants (Birt and Hurd 1976; Jennings 1976).

It hardly seems necessary to point out that both granulocrine and eccrine processes assume the direct participation of the secretory cells. A more radical view is held by Vasiliev (1969, 1971, 1972). From his studies of the floral nectaries in *Heracleum*, *Acer platanoides*, and other species, he has concluded that the immense variation in fine structure he encounters indicates that nectary cells do not function in this way. He theorizes that the sugar is released from the phloem and trickles through the free space between the nectary cells to the exterior. According to Vasiliev, the activity of the secretory cells is directed toward the elaboration of sugar-modifying enzymes or hormonal steroids.

MODELS OF NECTARY FUNCTION

Only a few studies have considered the coordinated activity of all parts of the nectary in the transport and secretion of sugars. Figier (1971) noted that in *Vicia* nectaries the subglandular cells with their thin cytoplasm and general paucity of organelles seemed well adapted for sugar transport. He observed strong acid phosphatase activity in the area of the plasmalemma, in ER cisternae and dictyosomes, and in some vacuoles as well as particularly intense phosphatase activity in the companion cells, especially in the area of the plasmalemma and in the small vacuoles. Figier suggested that sugars are actively transported from the sieve element by the companion cells and are then moved across the cell membranes of the subglandular and glandular tissue by a series of phosphorylations and dephosphorylations of the sugar molecules. The endoplasmic reticulum and dictyosomes were presumed to function as sites for synthesis and transport of the enzyme to the plasmalemma while the vacuoles, particularly those of the companion cells, were areas for sucrose conversion, since the nectar of *V. faba* contains some monosaccharides.

On the basis of the thorough studies of *Abutilon*, it has been postulated that active transport may occur at the level of the stalk cells in these nectaries (Findlay and Mercer 1971). As sugars are moved into the nectary hair by the activity of the stalk cell an osmotic gradient is created, water flows into the hair and the resulting increase in hydrostatic pressure forces the pre-nectar through the plasmalemma to the outside. The cutinization of the radial walls of the stalk cell, its strategic location at the base of the nectary hair, and numerous physiological studies on water potential in isolated nectaries suggest this method of sugar transport.

It is believed that phosphorylation and dephosphorylation mechanisms are probably not the method whereby such transport is accomplished since the available high energy phosphate is less than would be required for the observed rate of exudation. Gunning and Hughes (1976) have pointed out that the stalk cell with its relatively thin cytoplasm and reduced number of organelles does not appear to be suited to the membrane transport of sugars as envisioned by Reed et al. (1971). They have provided calculations that show the feasibility of a symplastic pathway involving bulk flow through the plasmodesmata with active transport of sugars over the entire surface of the nectary trichomes and sugar modifications possibly occurring enroute from the phloem. Compartmentalization within the ER was suggested but considered less attractive because of the complex network of cisternae associated with the plasmodesmata.

Fahn (1979a) has proposed a general model for nectar secretion in which sugars (pre-nectar) are transported through the symplast of the nectariferous tissue by means of the ER. (By nectariferous tissue Fahn apparently means either a specialized secretory parenchyma such as is present in *Citrus* or *Vinca* floral nectaries or, as in *Lonicera*, the layers of subglandular cells beneath the secretory hairs). Sugars then accumulate in the ER (or dictyosome vesicles) of the secretory cells and fusion of these membranes with the plasmalemma releases the nectar to the exterior.

Although the electron microscope evidence presented by many workers seems to support the argument that a granulocrine method of sugar secretion is operating in nectaries, some problems arise which need further investigation. For example, a variety of techniques have demonstrated beyond a doubt that dictyosomes vesicles may fuse with the plasmalemma, but the situation is less clear for cisternae and vesicles of ER, especially since the physical dimensions of the two structures are dissimilar. Using freeze-etch methods, Satir (1974) studied the fusion of mucocysts with the plasmalemma in *Tetrahymena* and *Paramecium*. She demonstrated that there is a rather characteristic redistribution of particles in the plasmalemma at the site where these vesicles will fuse. The evidence suggested an incorporation of ER membrane with that of the plasmalemma. Other studies of growing plant cells have provided some evidence that smooth ER, in continuity with rough cisternae, is able to fuse with the plasmalemma

(Morré and Van Der Woude 1974). The suggestion by Heinrich (1975b) that the actual fusion of the two membranes may be of such short duration that it could be easily missed in sections, further underlines the need for other kinds of microscopic evidence.

If dictyosomes and/or ER are involved in granulocrine secretion, there still remains the problem of how sugars (pre-nectar) enter these compartments. Fahn (1979b) has suggested that 1) they may be loaded directly into these organelles from the cytoplasm of the secretory cell after molecular transport from the phloem or that 2) they are localized within cisternae of ER throughout the nectary. In both cases, concentration of phloem sugars must occur either within these membranes or at the point where sugars are unloaded from the phloem.

With respect to the first proposal, it seems that if sugar molecules have been transported across nectariferous tissue membranes to the secretory cells, loading of these molecules into ER or dictyosomes prior to secretion is a superfluous and wasteful step unless it can be demonstrated that the ER is engaged in modification of these sugars. A similar transport of photosynthate into the ER of mesophyll cells was proposed in the early studies of phloem loading, but there is currently no evidence to support this idea (Geiger 1976). If, on the other hand, the ER serves as a channel through which sugars move from the phloem, one would expect large numbers of dilated cisternae to be a feature of the subglandular tissue which is not the case. It is also difficult to imagine how such cisternae, presumably carrying high concentrations of sugar, could avoid the same kinds of osmotic injury which protoplasts experience when subjected to hypotonic solutions.

Fahn and Rachmilevitz (1970) observed wall proliferations in *Lonicera* secretory cells. These reached their maximum development just prior to secretion and were associated with numerous cisternae of ER which reportedly fused with the plasmalemma. In this tissue a phenomenon (wall proliferation) heretofore associated with rapid short distance transport is presumably operating in conjunction with a granulocrine system. Again, it is difficult to show with certainty that such fusions are occurring, and the generation of so much wall material when secretion by vesicles would seem to be sufficient is difficult to reconcile. From a mechanical viewpoint, it seems that such protuberances would be a useless development unless all portions of the increased plasmalemma surface are readily accessible for vesicle fusion and this does not seem likely.

The presence of a heavily invaginated plasmalemma as evidence of excess membrane resulting from fusion of vesicles does not constitute a strong argument for granulocrine secretion since the appearance of invaginations often does not correlate with secretory activity (Findlay and Mercer 1960; Schnepf 1964c; Durkee et al, 1981).

The observation that the ratios of sugars in nectar exhibits considerable intraspecific constancy has been used as an argument for possible compartmentalization of sugar. It has been suggested that since enzymes of intermediary metabolism are abundant in nectary tissue cytoplasm, one might expect considerable fluctuations in these ratios unless the sugar solution is afforded some degree of protection from these enzymes (Gunning and Hughes 1976). Thus the ER and dictyosomes could serve as sequestering as well as transport agents. It seems unnecessary to invoke such a sequestering mechanism. If the enzymes of intermediary metabolism are behaving in nectaries as they do *in vitro*, their activity is governed by the principles of enzyme kinetics and various genetic and catalytic regulatory systems which will produce a secretion characteristic of the particular species.

The almost universal occurrence and abundance of ER in gland cells implies a role in secretory activity, a role which does not yet seem to be clearly defined.

THE FUTURE OF NECTARY RESEARCH

As must be evident by the conflicting and frequently inconclusive data, serious problems exist with the methods whereby nectaries have been studied. For example, there has been heavy emphasis placed upon the demonstration of phosphorylating enzymes in secretory cells. The reliability of the lead precipitation method has been questioned and it is now shown that the site of lead deposition may not accurately reflect the location of acid phosphatase within the cell (Washitani and Sato 1976). This inability to pinpoint the site of activity together with the ubiquity and lack of specificity of this enzyme (Lüttge and Schnepf 1976) suggests that care must be exercised in interpreting results using this method. Autoradiography carries its own set of problems, one of which is the current unavailability of a variety of suitably labeled compounds which are not readily incorporated into all parts of the cells. The drawback of using water-soluble tritiated sucrose has already been acknowledged (Fahn and Rachmilevitz 1975).

The great need for continuing ultrastructural studies of a variety of nectary types cannot be overemphasized. It would be particularly useful to concentrate our efforts on those species in which the nectar constituents have been characterized rather thoroughly. The extensive work of Baker and Baker (1975) would be a logical starting point in the selection of suitable material. In connection with this, the identification of specific amino acids in nectar may uncover unusual types which could be labeled and their path traced in the nectary tissue. Freeze-etch studies of those nectaries in which a granulocrine secretion mechanism has been proposed may finally resolve the ambiguous results which sometimes are obtained from conventional sectioning techniques.

More attention should be given to the nature and functioning of the subglandular tissue. Wergin et al. (1975) have already described some interesting arrangements of the plasmodesmata of this tissue in *Gossypium* which suggest that it functions in the symplastic transport of sugars from the phloem. In this context, we need to know more about the structure of nectary plasmodesmata and their relationship to the endoplasmic reticulum. Can they serve as valves regulating the intercellular movement of sugar solutions? Do the plasmodesmata always possess desmotubules or does the nature and presence of these structures change in response to variations in the activity of the tissue?

The vascular supply to the nectaries is especially worthy of study. The existence of a companion cell-sieve element complex in the nectary vascular bundle of *Passiflora* which is morphologically similar to that in leaf minor veins (Durkee 1977; Durkee et al. 1981) suggests that nectar vasculature would provide excellent material for investigation of phloem physiology, particularly phloem unloading.

Commonly used sampling methods may have to be revised. For example, refractometers are often employed to determine sugar concentrations in nectar, but there is now strong evidence that the non-sugar components of nectar, notably the amino acids, may significantly alter these readings (Inouye et al. 1980). Techniques which more nearly maintain the existing water potential of phloem sap during collection will also yield more accurate information about sugar content and make comparisons with nectar sugar concentration more meaningful. The aphid stylet technique is deservedly popular but personal experience has shown that aphids tend to be particular about their host plant and are virtually impossible to cultivate on many species. There needs to be more sampling and analysis of nectar and phloem sap from the *same* species. Most phloem data have been obtained from tree species (Zimmerman and Ziegler 1975) and a few favorable herbaceous types, but there is little overlap between these data and that available from nectar studies. If this information could be supplemented with measurements of sugar concentration at different levels within the nectary, or even within specific organelles, we could obtain a clearer picture of the secretion mechanism.

Ultimately, what is most needed are the combined and coordinated efforts of workers from a variety of disciplines. It is hoped that such cooperation will occur and stimulate further new and creative approaches in nectary research.

FIGURES

Material for transmission electron microscopy was fixed in 3 percent glutaraldehyde and post-fixed with 1 percent osmium tetroxide in sodium cacodylate buffer.

Unless otherwise indicated, the linear scale at the lower left corner of each photomicrograph indicates one micron.

Special thanks go to Dr. Gene Shih, Department of Botany, University of Iowa, for assisting with the scanning electron microscope and to Dr. John McDougal, Department of Botany, Duke University, for providing some of the plant material.

REFERENCES

Agthe, C. 1951. Über die physiologische Herkunst des Pflanzennektars. Ber. Schweiz. *Bot. Ges.* 61:240–273.

Baker, D. A., J. L. Hall, and J. R. Thorpe. 1978. A study of the extra-floral nectaries of *Ricinus communis. New Phytol.* 81:129–137.

Baker, H. and I. Baker. 1975. Studies of nectar-constitution and pollinator-plant coevolution. In L. E. Gilbert and P. H. Raven, eds., *Coevolution of Animals and Plants*, pp. 100–140. Austin: University of Texas press.

Bassot, J. M. 1966. Une forme microtubulaire et paracrystalline de reticulum endoplasmatique dans les photocytes des annelides polynoinae. *J. Cell. Biol.* 31:135–158.

Benner, U. and E. Schnepf. 1975. Die Morphologie der Nektarausscheidung bei Bromeliaceen: Beteiligung des Golgi-apparates. *Protoplasma* 85:337–349.

Birt, L. M. and F. J. R. Hird. 1958. Kinetic aspects of the uptake of amino acids by carrot tissue. *Biochem. J.* 70:286–292.

Brown, W. H. 1938. The bearing of nectaries on the phylogeny of flowering plants. *Proc. Amer. Phil. Soc.* 79:549–596.

Budnowski, A. 1922. Die Septaldrusen der Bromeliaceen. *Bot. Arch.* 1:47–80, 101–105.

Camp, R. R. and W. F. Whittingham. 1972. Host-parasite relationships in sooty blotch disease of white clover. *Amer. J. Bot.* 59:1057–1067.

Cusset, M. G. 1965. Les nectaires extra-floraux et la valeur de la feuille des Passifloracées. *Rev. Gen. Bot.* 72:145–219.

Durkee, L. T. 1977. The structure and function of extrafloral nectaries of *Passiflora*. Ph.D. dissertation, University of Iowa.

Durkee, L. T. In press. The floral and extrafloral nectaries of *Passiflora*. II. The extrafloral nectary. *Amer. J. Bot.*

Durkee, L. T., D. J. Gaal, and W. H. Reisner. 1981. The floral and extra-floral nectaries of *Passiflora*. I. The floral nectary. *Amer. J. Bot.* 68:453–462.

Esau, K. 1967. Minor veins in *Beta* leaves: structure related to function. *Proc. Amer. Phl. Soc.* 111:219–233.

Eymé, J. 1963. Observations cytologiques sur les nectaires de trois Ranunculacées. *Le Botaniste.* 46:137–179.

Eymé, J. 1966. Infrastructure des cellules nectarigènes de *Diplotaxus erucoides* D.C., *Helleborus niger* L., et *Helleborus foetidus* L. C. R. [Hebd. Seances] Acad. Sci. Ser. D. Nat. Sci. 262:1629–1632.

Eymé, J. 1967. Nouvelles observations sur l'infrastructure des tissues nectarigènes floraux. *Le Botaniste.* 50:169–183.

Eymé, J. and M. LeBlanc. 1963. Contribution à l'étude inframicroscopique d'inclusions cytoplasmiques présentes dans les ovules de *Ficaria* et dans les nectaires d'*Helleborus*. C. R. Hebd. Seances Acad. Sci. Ser. D. Sci. Nat. 256:4958–4959.

Fahn, A. 1952. On the structure of floral nectaries. *Bot. Gaz.* 113:464–470.

Fahn, A. 1979a. Ultrastructure of nectaries in relation to nectar secretion. *Am. J. Bot.* 66:977–985.

Fahn, A. 1979b. *Secretory Tissues in Plants.* New York: Academic Press.

Fahn, A. and P. Benouaiche. 1979. Ultrastructure, development, and secretion in the nectary of banana flowers. *Ann. Bot.* 44:85–93.

Fahn, A. and T. Rachmilevitz. 1970. Ultrastructure and nectar secretion in *Lonicera japonica*. In N. K. B. Robson, D. F. Cutler, and M. Gregory, eds., *New Research in Plant Anatomy*, pp. 51–56. London: Academic Press.

Fahn, A. and T. Rachmilevitz. 1975. An autoradiographical study of nectar secretion in *Lonicera japonica* Thunb. *Ann. Bot.* 39:975–976.

Figier, J. 1968. Localisation infrastructurale de la phosphomonoestérase acide dans la stipule de *Vicia faba* au niveau du nectaire. Rôle possible de cette enzyme dans les méchanismes de la sécrétion. *Planta.* 83:60–79.

Figier, J. 1971. Etude infrastructurale de la stipule de *Vicia faba* L., au niveau du nectaire. *Planta* 98:312–349.

Findlay, N. and F. V. Mercer. 1971. Nectar production in *Abutilon*. II. Submicroscopic structure of the nectary. *Aust. J. Biol. Sci.* 24:657–664.

Findlay, N., M. L. Reed, and F. V. Mercer. 1971. Nectar production in *Abutilon*. III. Sugar secretion. *Aust. J. Biol. Sci.* 24:665–675.

Frey-Wyssling, A. and C. Agthe. 1950. Nektar ist augeschiedener Phloemsaft. *Verhdlgn. Schweiz. Naturf. Ges.* 130:175.

Frey-Wyssling, A. and E. Hausermann. 1960. Deutung der gestaltlosen Nektarien. *Ber. Schweiz. Bot. Ges.* 70:150–162.

Geiger, D. R. 1966. Effect of sink region cooling on translocation of photosynthate. *Plant. Phys.* 41:1667–1672.

Giaquinta, R. T. 1979. Phloem loading of sucrose: Involvement of membrane ATPase and proton transport. *Plant Physiol.* 63:744–748.

Grout, B. W. W. and A. Williams. 1980. Extrafloral nectaries of *Dioscorea rotundata* Poir: Their structure and secretions. *Ann. Bot.* 46:255–258.

Gunning, B. E. S. and J. S. Pate. 1969. Transfer cells: Plant cells with wall ingrowths, specialized in relation to short distance transport of solutes—their occurrence, structure, and development. *Protoplasma* 68:107–133.

Gunning, B. E. S. and J. E. Hughes. 1976. Quantitative assessment of symplastic transport of pro-nectar into the trichomes of *Abutilon* nectaries. *Aust. J. Plant Physiol.* 3:619–637.

Heinrich, G. 1975a. Über den Glucose-metabolismus in Nektarien zweier Aloe-arten und über den Mechanismus der Pronektarsekretion. *Protoplasma* 85:351–371.

Heinrich, G. 1975b. Über die Lokalisation verschiedener Phosphatasen im Nektarium von *Aloe*. *Cytobiologie* 11:247–263.

Huber, H. 1956. Die Abhängigkeit der Nektarsekretion von Temperatur, Luft, und Bodenfeuchtigkeit. *Planta* 48:47–98.

Hutchings, V. M. 1978. Sucrose and proton co-transport in *Ricinus* cotyledons. I. H^+ influx associated with sucrose uptake. *Planta* 138:229–235.

Inouye, D. W., N. D. Favre, J. A. Lenum, D. M. Levine, J. B. Meyers, M. S. Roberts, F. T. Tsao, and Y. Wong. 1980. The effects of nonsugar nectar constituents on estimates of nectar energy content. *Ecology* 61:992–996.

Inouye, D. W. and R. S. Inouye. 1980. The amino acids of extrafloral nectar from *Helianthella quinquenervis* (Asteraceae). *Am. J. Bot.* 67:1394–1396.

Jennings, D. H. 1976. Transport in fungus cells. In *Encyclopedia of Plant Physiology*. New Series. *Transport in Plants*, vo.. 2, part A, U. Lüttge and M. G. Pitman, eds., *Cells*, pp. 189–228. New York: Springer.

Kennecke, M., H. Ziegler, and M. A. R. de Fekete. 1971. Enzymaktivitaten im Siebrohrensaft von *Robinia pseudoacacida* L. und anderen Baumarten. *Planta* 98:330–356.

Komor, E. 1977. Sucrose uptake by cotyledons of *Ricinus communis* L.: Characteristics, mechanisms, and regulation. *Planta* 137:119–131.

Luck, D. J. L. 1961. Glycogen synthesis from uridine diphosphate glucose and the distribution of the enzyme in liver cell fractions. *J. Biophys. Biochem. Cytol.* 10:195–209.

Lüttge, U. 1961. Über die Zusammensetzung des Nektars und den Mechanismus seiner Sekretion. I. *Planta* 56:189–212.

Lüttge, U. and E. Schnepf. 1976. Organic substances. In *Encyclopedia of Plant Physiology*. New series. *Transport in Plants*, vol. 2, part B, U. Lüttge and M. C. Pitman, eds., *Tissues and Organs*, pp. 244–277. New York: Springer.

Malek, F. and D. A. Baker. 1977. Proton co-transport of sugars in phloem loading. *Planta* 135:297–299.

Malek, F. and D. A. Baker. 1978. Effect of fusicoccin on proton co-transport of sugars in the phloem loading of *Ricinus communis* L. *Plant Sci. Letters* 11:233–239.

Mercer, F. V. and N. Rathgeber. 1962. Nectar secretion and cell membranes. *Proc. 5th Internatl. Congr. Electron Micros.*, vol. 2. WW-11. New York: Academic Press.

Morré, D. J., D. D. Jones, and H. H. Mollenhauer. 1967. Golgi apparatus mediated polysaccharide secretion by outer root cap cells of *Zea mays* I. Kinetics and secretory pathway. *Planta* 74:286–301.

Morré, D. J. and W. J. Van der Waude. 1974. Origin and growth of cell surface components. In E. D. Hay, T. J. King, and J. Papaconstantinou, eds., *Micromolecules Regulating Growth and Development*, pp. 81–111. 30th Symposium of the Society for Developmental Biology. New York: Academic Press.

Percival, M. S. 1961. Types of nectar in angiosperms. *New Phytol.* 60:235–281.

Pickett, C. H. and W. D. Clark. 1979. The function of extrafloral nectaries in *Opuntia acanthocarpa* (Cactaceae). *Am. J. Bot.* 66:618–625.

Porter, K. 1963. Diversity of the sub-cellular level and its significance. In J. M. Allen, ed., *The Nature of Biological Diversity*, pp. 121–163. New York: McGraw-Hill.

Rachmilevitz, T. and A. Fahn. 1973. Ultrastructure of nectaries of *Vinca Rosea* L., *Vinca major* L., and *Citrus sinensis* Osbeck., c.v. Valencia and its relation to the mechanism of nectar secretion. *Ann. Bot.* 37:1–11.

Rachmilevitz, T., and A. Fahn. 1975. The floral nectary of *Tropaeoleum majus*

L.—the nature of the secretory cells and the manner of nectar secretion. *Ann. Bot.* 39:721–728.

Racusen, R. H. and A. W. Galston. 1977. Electrical evidence for rhythmic changes in the co-transplant of sucrose and hydrogen ions in *Samanea pulvini. Planta* 135:57–62.

Satir, B. 1974. Membrane events during the secretory process. *Symp. Soc. Exp. Biol.* 28:399–415.

Schnepf, E. 1964a. Über Zellwandstrukturen bei den Köpfchendrusen der Schuppenblätter von *Lathraea clandestina* L. *Planta* 60:473–482.

Schnepf, E. 1964b. Zur Cytologie und Physiologie pflanzlicher Drusen. 4 Teil. Licht- und electronenmikroscopische Untersuchungen an Septalnektarien. *Protoplasma* 58:137–171.

Schnepf, E. 1964c. Zur Cytologie und Physiologie pflanzlicher Drusen. 5. Teil. Electronenmikroscopische Untersuchungen an Cyathialnektarien von *Euphorbia pulcherrima* in verschiedenen Funktionzustanden. *Protoplasma* 58:193–219.

Schnepf, E. 1968. Fine structure of mucus-secreting gland hairs on the ochrea of *Rumex* and *Rheum. Planta* 79:22–34.

Schnepf, E. 1969. Sekretion und Exkretion bei Pflanzen. *Protoplasmatologia*, vol. 8, no. 8. Vienna–New York: Springer.

Schnepf, E. 1973. Sezernierence und exzernierence Zellen bei Pflanzen. In G. C. Hirsh, H. Ruska, and P. Sitte, eds., *Grundlagen der Cytologie*, pp. 461–447. Stuttgart: Fischer.

Tacina, F. 1973. La structure optique et electronmicroscopique de la glande nectarifere chez *Cynoglossum officinale* L. *Rev. Roum. Biol.-Botanique* 14:201–209.

Vasiliev, A. E. 1969. Submicroscopic morphology of nectary cells and problems of nectar secretion. (In Russian). English abstract. Akad. Nauk. SSSR. *Bot. J.* 54:1015–1031.

Vasiliev, A. E. 1971. New data on the ultrastructure of flower nectaries. (In Russian). English abstract. Akad. Nauk. SSSR. *Bot. J.* 56:1292–1306.

Vasiliev, A. E. 1972. The ultrastructure of the nectary cells of cucumber. (In Russian). English abstract. Akad. Nauk. SSSR. *Tsitologiia* 14:405–415.

Vis, J. H. 1958. The histochemical demonstration of acid phosphatase in nectaries. *Acta Bot. Neerl.* 7:124–130.

Washatani, I. and S. Satro. 1976. On the reliability of the lead salt precipitation method of acid phosphatase localization in plant cells. *Protoplasma* 89:157–170.

Wergin, W. P., C. D. Elmore, B. W. Hanny, and B. Ingber. 1975. Ultrastructure of the subglandular cells form the foliar nectaries of cotton in relation to the distribution of plasmodesmata and the symplastic transport of nectar. *Amer. J. Bot.* 62:842–849.

Wooding, F. B. P. 1967. Endoplasmic reticulum aggregates of ordered structure. *Planta* 76:205–208.

Wrischer, M. 1962. Electronic microscope examination of the extrafloral nectaries of *Vicia faba. Acta Bot. Croat.* 20/21:75–94.

Zachariah, K. and R. H. Anderson. 1973. On the distribution and development of lattice bodies in apothecial cells of the fungus *Ascobolus. J. Ultrastructure Res.* 44:405–420.

Zandonella, P. 1970. Infrastructure des cellules du tissu nectarigens floral de quelques Caryophyllaceae. C. R. Acad. Sci. Paris, Series D. 270:1310–1313.

Zaurolov, O. A. and O. A. Pavlinova. 1974. Transport and conversion of sugars in nectaries in connection with the secretory function. (In Russian). *Tiziologiya Rastenii* 22:500–507.

Ziegler, H. 1956. Untersuchungen über die Leitung und Sekretion der Assimilate. *Planta* 47:447–500.

Zimmerman, M. 1952. Über ein neues Trisaccharid. *Experientia* 8:424–425.

Zimmerman, M. and H. Ziegler. 1975. List of sugars and sugar alcohols in sieve-tube exudates. In *Encyclopedia of Plant Physiology.* New Series. *Transport in Plants,* vol. 1, M. H. Zimmerman and J. A. Milburn, eds., *Phloem Transport,* pp. 480–503. New York: Springer.

2

NECTAR PRODUCTION IN A TROPICAL ECOSYSTEM

PAUL A. OPLER
U.S. FISH AND WILDLIFE SERVICE

In this paper I shall examine the role which nectar plays as an ecological mediator between one set of plants which produces it, and one set of animals which relies upon it for food.

The unifying theme is that all the plants and animals occur in one specific region, the lowland tropical dry forest life zone (Holdridge 1967) as found in northwestern Costa Rica. Herein, I describe the quantitative relationships and seasonal interplay between a selected sample of some 600 species of nectariferous plants (table 2.1) and the more than 1,200 species of nectarivorous animals found in that region (table 2.10).

THE STUDY AREA

The area where my studies were carried out, a 40 by 10 km area between Cañas and Liberia, Guanacaste Province, Costa Rica, lies at 85° west longitude and 10° north latitude between elevations of 25 and 200 m. The climate of the region has been treated in detail by Frankie et al. (1974) and Turner (1975). Daily temperatures vary little seasonally, and average 28°C, while rainfall averages 1600–1800 mm annually. Most rain falls between May 1 and November 15. Few rainfall events occur between November 15 and May 1, although those that do are important to the flowering of some plants (Opler et al. 1976). During the dry season, strong east to west trade winds blow and increase soil moisture deficits dramatically.

Within the area, land forms and soil types vary considerably. These differences, together with past and present land use practices by man, have produced a wide array of plant communities. Although I made an attempt to work in minimally disturbed habitats, it would be misleading to omit the fact that many sites were highly disturbed. Many of the plant species may have immigrated into the area in response to man's perturbations, a fact which demands caution when one attempts to draw coevolutionary generalities. Grazing, clearing, and fire have effected every habitat to one degree or another, and it is difficult to ascertain whether some particular kinds of habitat are natural or man-initiated.

Nine major communities may be recognized (see also Frankie et al. 1974), each possessing distinctive floral elements, although there exists wide variability between habitats within each: hillside deciduous forest, hillside evergreen forest, live oak (*Quercus oleoides*) forest, riparian evergreen forest, open pumice savanna, open flat savanna, seasonal swamps, pastures, and roadside edge.

Hillside deciduous forest was probably the most extensive original community, but now exists mainly as small isolated patches. It is characterized by a wide variety of deciduous trees, shrubs, and lianas, together with a few cacti and epiphytes. Only a few herbaceous plants, primarily perennial, occur in this community. Most common of the latter are several species of nonnectariferous *Dorstenia* (Moraceae).

Hillside evergreen forest occurs as rare patches within deciduous forest where high water table or seeps allow its occurrence. *Hymenaea courbaril* (Caesalpiniaceae), *Andira inermis* (Fabaceae), and *Ficus* spp. (Moraceae) usually constitute the overstory of these anomalous patches.

Oak forest, semi-deciduous, and dominated by *Quercus oleoides* (Fagaceae), occurs on deep pumice soils. Today, there are probably no mature examples extant in the area. The areas studied contained few tree species, but a fairly rich assortment of shrubs, lianas, and vines. Only a few herbaceous species are present.

Riparian evergreen forests occur along all major and minor streams of the area and are important dry season refuges for many animals of the area, including insects (Janzen 1973). Floristically the richest forest habitat, it contains the widest array of tree, shrub, and liana species, but almost no herbs, except on recently created sand or gravel bars.

Open pumice savannas, occurring on material produced by Volcan Miravalles, a few miles to the northwest, occur as grasslands, with scattered trees of *Quercus oleoides*, *Byrsonima crassifolia* (Malpighiaceae), and *Curatella americana* (Dilleniaceae). A wide array of distinctive herbs and perennial subshrubs are scattered throughout this habitat. Although this community is obviously a fire-maintained disclimax, I feel that it is a natural one and not man-initiated as is indicated by the preponderance of native plants and distinct lack of pan-tropical or pan-american "weeds."

Open flat savanna, often appearing more like high African plains, are probably man-initiated and man-maintained. These areas, now the most widespread local habitat, are created by almost complete felling, and then maintained by heavy cattle grazing, fire, and machete. The widely scattered trees are a miscellaneous assortment of deciduous forest species, but are often dominated by several legumes and by several *Tabebuia* (Bignoniaceae). The shrubs present are usually those ignored by cattle, and the herbaceous layer consists of many grasses and herbs which thrive on disturbance.

Seasonal swamps are flat areas shallowly flooded with water for long episodes during the wet season. Dominated by few trees such as *Crescentia alata* (Bignoniaceae), *Guazuma ulmifolia* (Sterculiaceae), *Coccoloba caracasana* (Polygonaceae), or *Parkinsonia aculeata* (Caesalpiniaceae), these habitats are richest in epiphytes, and also host a wealth of herbs, vines, and lianas.

Pastures are more intensively maintained and grazed than the open flat savanna habitats—which they resemble floristically, although they may be irrigated, they are often treeless, and harbor many pan-tropical and pan-american weeds.

Roadsides, especially the gravel-edged Pan-American Highway, are heavily trodden by people and livestock and are frequently burnt or cut. Receiving a varied seed input from passing traffic, many pioneer plants were found as isolated individuals or patches only in this habitat.

THE PLANTS

In introducing my "cast" of 587 nectariferous plants (table 2.1), I should point out that it probably represents the majority of such plants in the area, but by no means all. The list excludes those taxa not producing floral nectar, e.g., Piperaceae, Moraceae, Solanaceae (except *Cestrum*), many Caesalpiniaceae (*Cassia*), Bixaceae, Cochlospermaceae, Poaceae, and Cyperaceae. Taxa which may or may not produce floral nectar were included, as were the Malpighiaceae most which produce nectar on the back of calyces.

Ten families were dominant among the nectariferous plants of the region, each with fifteen or more species (table 2.1): Fabaceae (78), Asteraceae (41), Bignoniaceae (36), Rubiaceae (28), Euphorbiaceae (27), Malvaceae (24), Convolvulaceae (22), Mimosaceae (21), Acanthaceae (19), and Labiatae (18). These families account for 314 (53 percent) of the 587 taxa studied, while the remaining 274 species are spread amongst eighty other families.

The family Fabaceae is dominant amongst trees, especially those of hillside deciduous forest, and is the most common family represented by herbaceous vines, and the second most common amongst herbs (table 2.1). The fabaceous vines and herbs are especially prolific in disturbed situations.

Asteraceae is the dominant family amongst herbs, with only a few shrubs and vines representing other life forms. Most family members are found in second growth habitats.

The Bignoniaceae are dominant amongst woody lianas, being found in all but the most open habitats. *Tabebuia neochrysantha*, one of six bignoniaceous trees, is dominant in many hillside forest habitats (Frankie et al. 1974). The floral and ecological aspects of this family have been studied by Gentry (1974, 1976), who included the same Guanacaste forests in his research.

The Rubiaceae were dominant amongst woody forest shrubs primarily due to *Psychotria* (6 species) and *Randia* (3 species). Most family members are native to the region, and almost all are found in forest habitats.

TABLE 2.1.
FLORAL NECTAR-PRODUCING PLANTS OF GUANACASTE
PROVINCE, COSTA RICA

Order – Family	Numbers of Species Representing Each Life Form				
	Epiphyte	*Herb*	*Shrub*	*Vine*	*Tree*
Magnoliales					
Lauraceae					1
Hernandiaceae					1
Nymphaeales					
Nymphaeaceae		2			
Caryophyllales					
Phytolaccaceae		2			
Nyctaginaceae		2		1	1
Cactaceae	2		3		1
Aizoaceae		1			
Portulacaceae		1			
Polygonales					
Polygonaceae					5
Dilleniales					
Dilleniaceae				2	1
Theales					
Ochnaceae			2		
Guttiferae					2
Malvales					
Elaeocarpaceae					2
Tiliaceae			1		4
Sterculiaceae		2	4		1
Bombacaceae					5
Malvaceae		17	7		
Violales					
Flacourtiaceae			2		5
Turneraceae		4			
Passifloraceae				6	
Caricaceae					1
Violaceae		1			
Begoniaceae		2			
Cucurbitaceae				7	
Capparales					
Capparidaceae		2	1		3
Ebenales					
Sapotaceae					2
Ebenaceae					1
Styracaceae					1
Primulales					
Theophrastaceae			1		
Myrsinaceae			2		

TABLE 2.1. Continued

Order – Family	Epiphyte	Herb	Shrub	Vine	Tree
Rosales					
Chrysobalanaceae			2		1
Mimosaceae		3	3	2	13
Caesalpiniaceae			1	1	7
Fabaceae		33	3	25	17
Myrtales					
Lythraceae		3	1		
Myrtaceae			2		2
Onagraceae		3			
Combretaceae				2	
Proteales					
Proteaceae					1
Santalales					
Olacaceae					2
Opiliaceae					1
Loranthaceae	1				
Celastrales					
Hippocrateaceae				1	1
Celastraceae					1
Euphorbiales					
Euphorbiaceae		15	8		4
Rhamnales					
Rhamnaceae				2	2
Vitaceae				3	
Sapindales					
Sapindaceae			2	6	2
Burseraceae					2
Anacardiaceae					5
Simaroubaceae			1		3
Rutaceae					1
Meliaceae					7
Zygophyllaceae		2			1
Geraniales					
Oxalidaceae		1			
Linales					
Erthroxylaceae			1		1
Polygalales					
Polygalaceae			3	1	
Malpighiaceae			2	7	1
Trigoniaceae				1	
Krameriaceae		1			
Umbellales					
Araliaceae					1

Numbers of Species Representing Each Life Form

TABLE 2.1. Continued

Order – Family	Epiphyte	Herb	Shrub	Vine	Tree
Gentianales					
Loganiaceae		1			
Gentianaceae		2			
Apocynaceae			3	4	3
Asclepiadaceae		2		8	
Polemoniales					
Solanaceae			3		
Convolvulaceae		4		18	
Cuscutaceae	1				
Lamiales					
Boraginaceae		2	4		5
Verbenaceae		4	4		1
Labiatae		18			
Scrophulariales					
Scrophulariaceae		3	1		
Gesneriaceae		3			
Bignoniaceae				30	6
Acanthaceae		18	1		
Campanulales					
Lobeliaceae		2			
Rubiales					
Rubiaceae		8	13		7
Asterales					
Asteraceae		34	5	2	
Alismatales					
Alismataceae		1			
Commelinales					
Commelinaceae		7			
Bromeliales					
Bromeliaceae		1			
Zingiberales					
Heliconiaceae		1			
Marantaceae		2			
Arecales					
Palmae			2		1
Liliales					
Pontederiaceae		2			
Amaryllidaceae		1			
Iridaceae		3			
Smilacaceae				1	
Dioscoreaceae				4	
Orchidales					
Orchidaceae	6	3			
TOTAL	10	219	88	135	136

Representation of nectariferous plants in Guanacaste Province, Costa Rica, listed by order, family, and life form. (Order of families based upon the Cronquist modification of the Tahktajan classificatory system.)

The Euphorbiaceae are about equally divided between herbs of highly disturbed habitats—especially *Chamaesyce* spp.—and shrubs and trees of forest habitats. With one exception, the remaining five families of the "big ten" (Malvaceae, Convolvulaceae, Mimosaceae, Acanthaceae, and Labiatae) are composed of herbaceous or slightly woody second growth herbs or vines. The Mimosaceae, although represented by several second growth pioneers, is dominated in the area by forest trees, especially species of *Albizzia* and *Pithecellobium*.

FLORAL ADAPTATIONS TO POLLINATORS

Flowering plants usually have their flowers adapted to take maximal advantage of one or more classes of pollinatory agents, whether they be physical or biotic. Floral morphology and behavior have evolved in combination to maximize pollinator attraction— when necessary, to increase likelihood of suitable fertilization, and to prevent visits by unefficacious pollinators.

Particular combinations of floral morphology and behavior, often referred to as "syndromes," have been reported as typifying plants serviced by different pollinatory groups (Faegri and Pijl 1966; Baker and Hurd 1968; Percival 1965; Proctor and Yeo 1973). Corolla color and shape, positioning of sexual parts, position on plant, presence of nectar guides, fragrances, and time of anthesis are employed most often. In this volume, Baker and Baker have explored the characteristics of nectar sugars typical of flowers visited by different pollinator classes, while Cruden et al. (see paper 3) explore timing and duration of nectar production for flowers with different pollinators.

At the onset it should be called to the reader's attention that no two plants have flowers exactly alike, and since plants of the same "syndrome," e.g., bat-pollinated, may belong to quite different phylogenetic lines, they may be quite dissimilar in appearance. In fact, flowers of a given "syndrome" often lack one or more of the expected features. Similarly, plants with flowers appearing to fit a given pollinatory group may in fact be pollinated by quite a different group. Furthermore, many plants are catholic in their pollinatory adaptations, being serviced by more than one of the pollinatory classes.

It is my purpose here to draw particuliar attention to the role which nectar *quantity* plays in these adaptations, and to relate it to several pollinatory classes.

Based upon a familiarity with the plant and pollinator communities of the Costa Rica study site (e.g., Frankie 1975), an effort was made to quantify nectar production,[1] floral biomass,[2] and flower length for at least ten plants in each of seven pollinatory groupings, i.e., bat, hummingbird, hawkmoth, medium/large bee, butterfly, settling moth, and wasp/small bee. In several cases, assignment to a class was somewhat arbitrary, usually being based not only on "syndrome" features, but upon which animal group was the most probable pollinator as indicated by our observations. A detailed study emphasizing tree pollination for this site is nearing completion (Haber, Frankie, Opler, unpub. data).

Nectar volume, floral biomass, and corolla length are presented for the plants representing each pollinatory class in tables 2.4–2.9, while a summary of this information is shown in table 2.2. Drawings of some representative flowers from each pollinatory grouping are shown in figures 2.2–2.6, 2.8–2.12, 2.14, 2.16–2.19, 2.21–2.23, 2.25, and are intended not only to document similarities of plants within "syndromes" but to document the range of variability.

NECTAR VOLUME AND FLORAL BIOMASS

Nectar volume is an indirect measure of the quantity of food reward offered by flowers to their animal visitors, the absolute quantity and quality being the price paid by the plant to attract appropriate pollinators (Heinrich 1975). The important contribution of sugar concentration is discussed elsewhere in this volume by the Bakers and Cruden et al. (papers 3 and 4).

As one considers plants adapted to increasingly large pollinators, one finds that the quantity of floral nectar produced increases in parallel—not an altogether surprising result (tables 2.2, 2.4–2.9). From

1. Nectar production was measured several times during peak flow using micropipettes of known volume to extract nectar from flowers which had been bagged prior to anthesis, thus preventing removal of nectar by animals or its evaporation. Amount of nectar varied considerably between flowers, and the largest amount found in a single flower was used to represent each species and is referred to herein as "maximum nectar." Nectar volumes are expressed in microliters (μl) in the tables and figures. Note that Cruden et al. (see paper 3) express their data in amount of sugar, i.e., sugar concentration multiplied by volume.
2. Floral biomass, expressed in grams (g), was obtained by weighing freshly picked flowers (distal to pedicel) after removal of nectar.

the smallest amount of nectar produced (0.03 μl by *Anacardium excel-sum*)[3] to the largest (9400 μl by *Ochroma lagopus*), a fivefold increase in order of magnitude is involved. Flowers adapted by wasps and small bees averaged only 0.63 μl maximum nectar,[4] while those adapted for bat pollination averaged 1310.5 μl. Plants with other pollinatory adaptations fell between these extremes. It should be noted that the mean value of all plants sampled in a group is unrepresentative on two counts: 1) the inordinately large effect on a small sample size of *Ochroma lagopus;* and 2) the fact that the plants sampled are biased toward the larger flowered representatives of the community. A "modal plant" in the community would probably have somewhat less than one microliter of maximum available nectar.

When one considers floral biomass (tables 2.2, 2.4–2.9), the same trend shown by nectar quantity is apparent, i.e., the larger the pollinator class, the heavier the flower. This size relationship is due at least in part to the fact that the nectaries that produce large amounts of nectar are usually larger in size. Of greater contribution to the larger sizes are the proportionately larger corollas of flowers adapted to large pollinators. Other possible contributing factors are larger or greater ovule number and the proportionately larger sepals of increasingly large flowers. Between the lightest (0.003g) and the heaviest (90g), there exists a greater than fourfold order of magnitude.

Mean corolla length (cm) of flowers adapted to the various visitor groups ranges from 0.41 cm (wasp/small bee flowers) to 6.99 cm (hawkmoth flowers), while the mean corolla length of all flowers measured is 0.20 cm (table 2.2). The order of mean corolla length values differs from both that of mean maximum nectar values, and that of mean floral biomass values (table 2.2).

The frequency distribution of corolla lengths is log-normal (fig. 2.29), while the frequency class represented by the largest number of species is 0.5 to 1.0 cm.

3. This was the smallest amount measurable by micropipettes. The florets of some small Asteraceae (Compositae) were so narrow that the nectar within was not measurable by my methods.
4. Most plants in this category are visited by a wide array of Hymenoptera, Lepidoptera, Diptera, and Coleoptera. The term "generalized" would be a better descriptor of these flower visitor assemblages, but it is felt that wasps and small bees are the most efficient and usual pollinators.

TABLE 2.2
NECTAR AND FLORAL FEATURES OF PLANTS ADAPTED TO DIFFERENT POLLINATORY TYPES

Pollinatory Adaptation	Maximum Available Nectar (μl)			Floral Biomass (g)			Corolla Length (cm)			$\bar{X}$ Flower Biomass / $\bar{X}$ Nectar (g/μl)	$\bar{X}$ Corolla / $\bar{X}$ Nectar (cm/μl)	$\bar{X}$ Nectar / $\bar{X}$ Visitor Biomass (g/μl)
	$\bar{X}$	S.D.	N	$\bar{X}$	S.D.	N	$\bar{X}$	S.D.	N			
Bat	1310.50	2938.30	10	13.39	28.03	10	6.06	3.14	10	0.01	0.005	0.02
Hummingbird	16.80	18.57	11	0.32	0.35	11	3.08	1.06	11	0.02	0.18	0.24
Hawkmoth	130.55	212.32	11	2.15	3.16	11	6.99	6.02	11	0.02	0.05	0.01
Large bee	9.75	18.96	19	0.60	0.84	19	3.05	2.04	19	0.06	0.31	0.05
Butterfly	0.93	0.86	8	0.02	0.015	8	1.34	0.72	8	0.02	1.44	0.19
Settling moth	0.84	0.60	6	0.02	0.014	3	1.15	0.62	6	0.02	1.37	–
Wasp/Small bee	0.63	0.68	14	0.02	0.019	13	0.41	0.16	14	0.03	0.65	0.16
All plants	184.36	1076.90	79	2.31	10.80	75	3.20	3.48	79	0.01	0.02	0.02

Nectar production and floral features of plants adapted to different pollinator groups.

REGRESSION ANALYSES

My study shows that floral biomass (g) and maximum available nectar (µl) are highly correlated. Linear regression analysis of the untransformed variables gives a correlation coefficient (r) of 0.99 (T = 845.46, p < 0.001) (table 2.3). When the data are logarithmically transformed, a lower but still significant degree of correlation (r = 0.87, T = 15.10, p < 0.001) is evidenced (table 2.3).

When the variables for plants in each pollinatory grouping are subjected to individual regression analyses, both normal and logarithmically transformed, most show a moderate to high degree of correlation (table 2.3, figs. 2.1, 2.7, 2.13, 2.15, 2.20, 2.24, 2.26). In all but one instance (i.e., wasp/small bee pollinated flowers), regression of the untransformed data gives a higher degree of correlation.

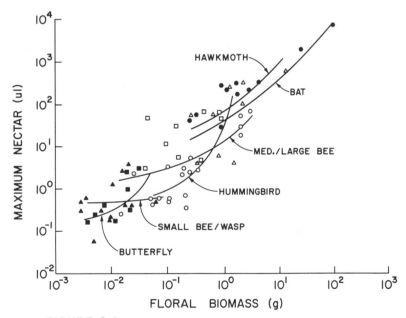

FIGURE 2.1

Linear regressions (on log-log scale) for plants of various pollinatory syndromes (settling moth plants omitted): floral biomass (g) vs. maximum nectar (µl). ● = bat-pollinated plants, △ = Hawkmoth-pollinated plants,□ = hummingbird-pollinated plants, ○ = medium to large bee-pollinated plants,■ = butterfly-pollinated plants, ▲ = small bee and wasp-pollinated plants.

TABLE 2.3

REGRESSION ANALYSIS (LINEAR AND LOGARITHMIC): FLORAL BIOMASS (g) VS. MAXIMUM NECTAR (µl) FOR PLANTS ADAPTED TO DIFFERENT POLLINATOR GROUPS

Pollinator Adaptation	Linear					Logarithmic				
	A_0	A_1	r	N	T	A_0	A_1	r	N	T
Bat	−88.61	104.48	.99	10	28.14*	2.04	0.83	.87	10	5.02*
Hummingbird	5.31	35.91	.67	11	2.71	1.40	0.77	.55	11	1.96
Hawkmoth	−6.02	63.50	.95	11	9.00*	1.51	1.02	.67	11	2.72
Large bee	1.89	19.23	.87	19	7.35*	1.08	1.03	.85	19	6.73*
Butterfly	−0.09	50.23	.89	8	4.75*	1.80	1.13	.82	8	3.52*
Settling moth	0.04	17.58	.52	3	0.71	−0.64	0.06	.04	3	0.04
Wasp/Small bee	0.64	1.92	.05	13	0.18	0.02	0.20	.14	13	0.47
All plants	−38.39	103.0	.99	75	845.*	1.48	1.03	.87	76	15.16*

Regression analysis (linear and logarithmic): Floral biomass (g) vs. maximum nectar (µl) for plants adapted to different pollinator groups.

A_0 = Y intercept
A_1 = regression coefficient
r = correlation coefficient
T = Student's T-value
* = p ⩽ 0.01

Only for two pollinator groups are the correlation coefficients so low that any degree of correlation must be rejected. For settling moth flowers, the effect of small sample size must be cited as the probable cause of poor correlation, while such cannot be the case for wasp/small bee flowers. Rather, the relatively small perianths of flowers adapted to the latter pollinatory group are overshadowed by the highly variable sized ovaries of the plants which constitute this group.

Bat flowers show the highest degree of correlation (r = 0.99, T = 28.14, p < 0.001), *Ochroma lagopus* having a large effect on the regression (table 2.3, fig. 2.7). Nevertheless, there are no plants which do not fit the regression well.

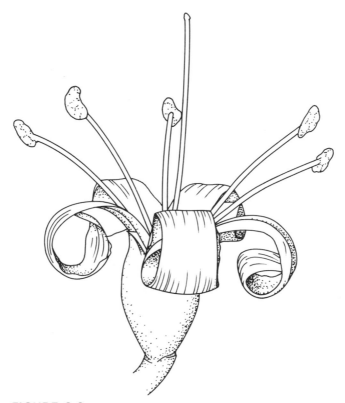

FIGURE 2.2
Ceiba acuminata (Bombacaceae), a bat-pollinated tree. Nectar accumulates at base of corolla.

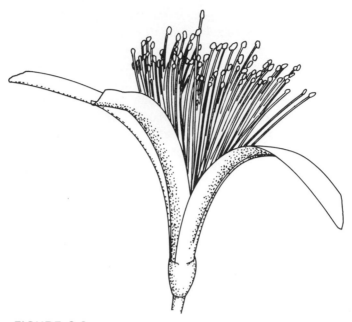

FIGURE 2.3
Bombacopsis quinata (Bombacaceae), a bat-pollinated tree. Nectar accumulates at base of corolla.

Hummingbird flowers only show moderate correlation in the regression analysis (table 2.3, fig. 2.13). This may be in part explained
by the suspicion that *Russellia sarmentosa* and *Quamoclit coccinea*,
both with red tubular flowers, may be more properly classed as butterfly pollinated, since butterflies were also frequent visitors.

Hawkmoth flowers show a degree of correlation in the regression
analyses second only to that of bat flowers in its significance (table
2.3, fig. 2.15). *Luehea candida*, which has the greatest effect on the
regression, does not present the general aspect of a "hawkmoth
flower" in some of its outward features. Its lack of a tubular corolla,
its broad stout stigma, and feeding stamens suggest either a "bat
flower" or a "beetle flower," however, Heithaus et al. (1975) did not
find its pollen on bats, nor did I observe bats during many hours of
observation. Although beetles were often observed feeding on the
stamens, they contacted the stigmata infrequently. The white petals,
fragrant odor, and copious nectar of this nocturnal flower, together

FIGURE 2.4

Crescentia alata (Bignoniaceae), a bat-pollinated tree. Nectar accumulates inside corolla.

TABLE 2.4
NECTAR PRODUCTION AND FLORAL FEATURES OF
NECTARIFEROUS PLANTS ADAPTED PRINCIPALLY FOR
BAT POLLINATION

Species	Family	Floral Biomass (g)	Corolla Length (cm)	Maximum Nectar (μl)
Ochroma lagopus	Bombacaceae	90.0	11.5	9400
Ceiba acuminata	Bombacaceae	26.1	6.3	2500
Hymenaea courbaril	Fabaceae	1.5	2.6	362
Bauhinia ungulata	Caesalpiniaceae	0.9	5.0	223
Ceiba pentandra	Bombacaceae	4.7	4.7	220
Bombacopsis quinata	Bombacaceae	7.2	8.6	125
Crescentia alata	Bignoniaceae	1.9	5.5	110
Crataeva tapia	Capparidaceae	0.4	1.2	65
Inga vera var. *spuria*	Mimosaceae	0.3	5.5	65
Bauhinia pauletia	Caesalpiniaceae	0.9	9.7	35
MEAN ($\bar{X}$)		13.39	6.06	1310.50

Nectar production and floral features of nectariferous plants adapted principally for bat pollination.

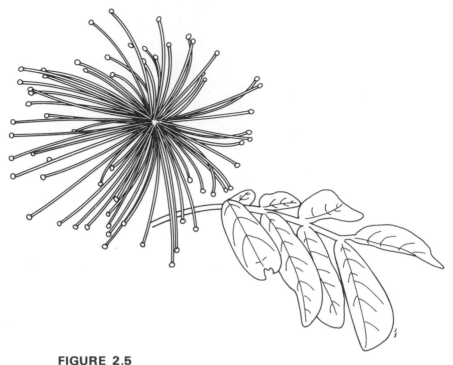

FIGURE 2.5

Inga vera var. *spuria* (Mimosaceae), a bat-pollinated tree. Nectar accumulates inside corolla.

with more recent observations by Haber and Frankie (in preparation), demonstrate that *L. candida* is, in fact, hawkmoth pollinated.

Medium to large bee-pollinated flowers showed a highly significant correlation between floral biomass and maximum nectar (r = 0.87, T = 7.35, p < 0.001), but with some larger flowered species not adhering to the regression line (table 2.3, fig. 2.20). *Thevetia ovata*, also visited and probably pollinated by large hawkmoths and hummingbirds, produced more nectar than its biomass would dictate for a bee flower, while *Centrosema plumieri* (see fig. 2.16) and *Tabebuia rosea* are also visited and perhaps pollinated by hummingbirds, as well as bees.

Butterfly-pollinated flowers showed very good correlation between the two variables (r = 0.89, T = 4.75, p < 0.001), with only *Justicea* sp. being far off the regression line (table 2.3, fig. 2.24).

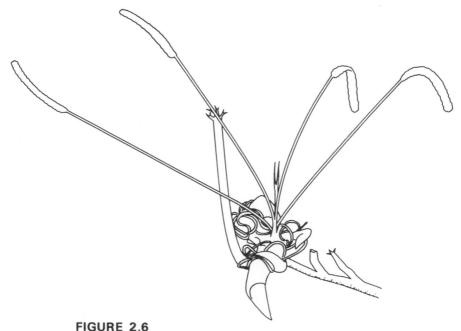

FIGURE 2.6
Bauhinia pauletia (Caesalpinaceae), a bat-pollinated shrub. Nectar accumulates in calyx.

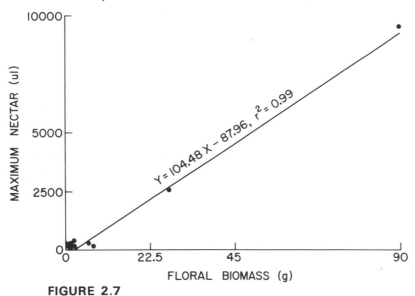

FIGURE 2.7
Linear regression of bat-pollinated plants: floral biomass (g) vs. maximum nectar (μl). Each point represents the values for a single species (see also table 2.4).

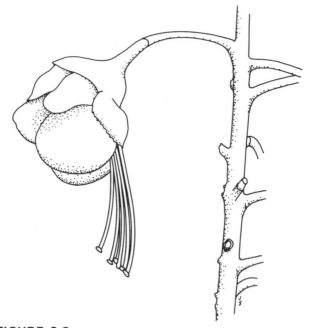

FIGURE 2.8
Caesalpinia conzattii (Caesalpinaceae), a hummingbird-pollinated tree.
Nectar accumulates inside corolla.

Settling moth flowers had a low insignificant correlation between the two variables ($r = 0.52$, $T = 0.71$, n.s.), as was previously explained. A more complete sampling of such flowers might have given a differ-ent result.

The poorest correlation between the variables was for wasp/small bee flowers ($r = 0.05$, $T = 0.18$, n.s.). The data for floral biomass and maximum nectar seem enigmatic for this pollinatory group (tables 2.3, 2.9, fig. 2.26), since there seem to be two trends or "strategies." One group seems to produce a similar amount of nectar (± 0.25 μl), but to have variable flower weight while the other produces variable amounts of nectar at a similar flower weight (0.01–0.02g). There is no apparent biological explanation for this pattern, and it may well be artifactual.

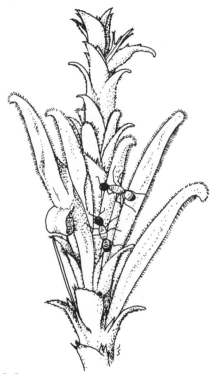

FIGURE 2.9

Aphelandra deppeana (Acanthaceae), a hummingbird-pollinated shrub. Nectar accumulates inside corolla. Note ants which tend extra-floral nectaries.

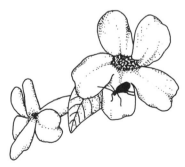

FIGURE 2.10

Ruellia inundata (Acanthaceae), a hummingbird-pollinated herb. Nectar accumulates inside corolla. Note chrysomelid beetle on flower is not a nectarivore.

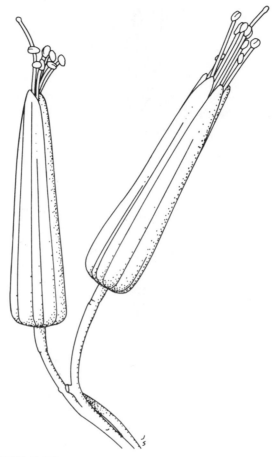

FIGURE 2.11

Quassia amara (Simaroubaceae), a hummingbird-pollinated shrub. Nectar accumulates inside corolla.

NECTAR SEASONALITY

The seasonality of nectar availability in any ecosystem may be viewed from any of a number of perspectives. One might choose to view seasonality from the standpoint of particular plant life forms, as has been done for Guanacaste trees by Frankie et al. (1974) and Frankie (1975), or one might choose to view nectar seasonality from

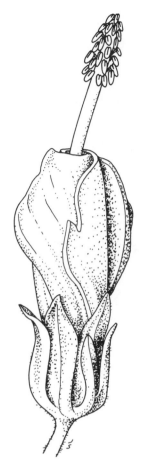

FIGURE 2.12
Malvaviscus arboreas (Malvaceae), a hummingbird-pollinated shrub.
Nectar accumulates in calyx and between petal bases.

the view of resource availability for particular animal groups or species, as has been done for bats by Heithaus et al. (1975). Furthermore, one might elect to examine nectar seasonality for particular plant taxa, as has been done for the genus *Cordia* in Guanacaste by Opler et al. (1977), for plants of particular breeding systems (Bawa and Opler 1975), or for plants occurring in different habitat types.

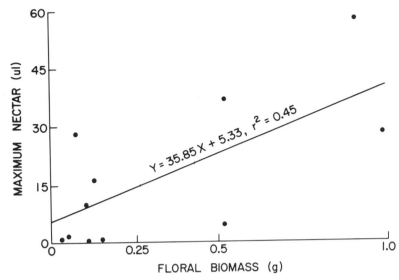

FIGURE 2.13

Linear regression of hummingbird-pollinated plants: floral biomass (g) vs. maximum nectar (μl). Each point represents the values for a single species (see also table 2.5).

TABLE 2.5

NECTAR PRODUCTION AND FLORAL FEATURES OF NECTARIFEROUS PLANTS ADAPTED PRINCIPALLY FOR HUMMINGBIRD POLLINATION

Species	Family	Floral Biomass (g)	Corolla Length (cm)	Maximum Nectar (μl)
Quassia amara	Simaroubaceae	0.87	4.0	56.0
Malvaviscus arboreas	Malvaceae	0.53	3.0	39.0
Combretum farinosum	Combretaceae	0.07	1.0	28.1
Caesalpinia conzattii	Caesalpiniaceae	0.98	3.7	26.6
Helicteres guazumaefolia	Sterculiaceae	0.12	2.6	16.6
Aphelandra deppeana	Acanthaceae	0.10	3.9	10.0
Struthanthus oerstedii	Loranthaceae	0.53	4.5	4.1
Ruellia inundata	Acanthaceae	0.04	3.4	1.8
Russellia sarmentosa	Scrophulariaceae	0.03	1.5	1.6
Quamoclit coccinea	Convolvulaceae	0.10	3.3	0.5
Hamelia patens	Rubiaceae	0.15	3.0	0.5
MEAN ($\overline{X}$)		0.32	3.08	16.80

Nectar production and floral features of nectariferous plants adapted principally for hummingbird pollination.

FIGURE 2.14
Hymenocallis littoralis (Amaryllidaceae), a hawkmoth-pollinated herb.
Nectar fills narrow portion of corolla tube.

Each viewpoint of nectar seasonality may allow different sorts of generalizations about the ecosystem, but I am concerned here with only a few of these perspectives. In particular, overall seasonality of nectar availability, variation and patchiness of nectar availability within and between habitat types, and lastly, seasonal availability for different pollinatory groups.

Considering the seasonality of all nectariferous plants in the ecosystem (fig. 2.27) one finds that the preponderance of species flower

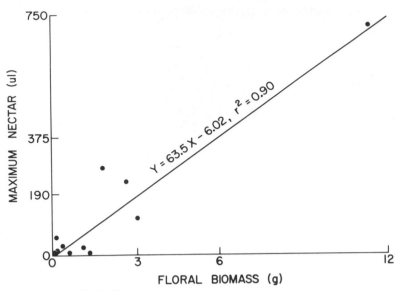

FIGURE 2.15

Linear regression of hawkmoth-pollinated plants: floral biomass (g) vs. maximum nectar (μl). Each point represents the values for a single species (see also table 2.6).

TABLE 2.6

NECTAR PRODUCTION AND FLORAL FEATURES OF NECTARIFEROUS PLANTS ADAPTED PRINCIPALLY FOR HAWKMOTH POLLINATION

Species	Family	Floral Biomass (g)	Corolla Length (cm)	Maximum Available Nectar (μl)
Luehea candida	Tiliaceae	11.3	4.1	710
Lindenia rivalis	Rubiaceae	1.7	17.2	264
Hymenocallis littoralis	Amaryllidaceae	2.5	19.0	220
Luehea speciosa	Tiliaceae	3.0	3.1	100
Isotoma sp.	Lobeliaceae	0.8	10.5	65
Albizzia longipedata	Mimosaceae	0.32	3.5	35
Randia subcordata (Pistillate)	Rubiaceae	0.9	6.5	23
Cordia gerascanthus	Boraginaceae	0.9	2.5	7
Pithecellobium saman	Mimosaceae	0.09	2.0	6.5
Randia subcordata (Staminate)	Rubiaceae	0.9	6.0	3.4
Alibertia edulis	Rubiaceae	1.25	2.5	2.2
MEAN ($\overline{X}$)		2.15	6.99	130.53

Nectar production and floral features of nectariferous plants adapted principally for hawkmoth pollination.

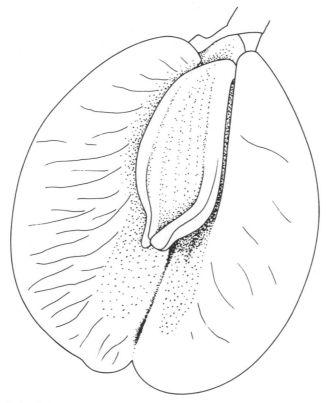

FIGURE 2.16
Centrosema sagittatum (Fabaceae), a large bee-pollinated herbaceous vine (closely allied to *C. plumieri*). Nectar accumulates inside corolla.

in the wet season (May through mid-November), while the fewest species flower during the dry season (mid-November through April). The contribution of trees to this overall pattern has been detailed by Frankie et al. (1974), while the separate contributions of shrubs, lianas, vines, and herbs are presently being summarized (Opler, Frankie, and Baker, unpublished). The overall preponderance of wet season flowering (fig. 2.27) is due largely to the contribution of herbs and vines, particularly those of disturbed habitats. This overall pattern may have been quite different prior to human disturbance. In fact,

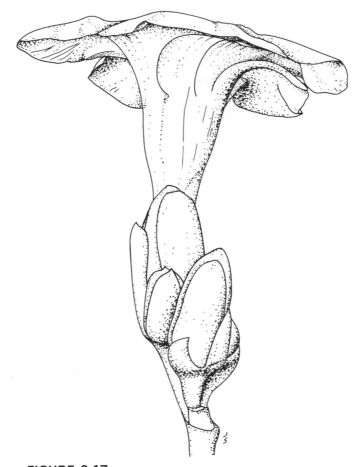

FIGURE 2.17
Stemmadenia obovata (Apocynaceae), a medium to large bee-pollinated tree. Nectar accumulates inside corolla.

amongst clearly indigenous plants, an overall peak of flowering oc-curs in the early wet season (May to July).

The different habitats in Guanacaste (see introductory discussion) present differing combinations of soil moisture availability, soil types, exposures, and light availability due to differing densities, heights, and leaf persistence of trees present. Two extremes in Guanacaste are

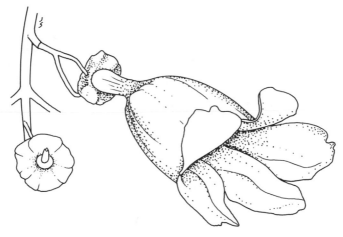

FIGURE 2.18

Petastoma patelliferum (Bignoniaceae), a large bee-pollinated liana. Nectar accumulates at base of corolla.

FIGURE 2.19

Myrospermum frutescens (Fabaceae), a medium to large bee-pollinated tree. Nectar accumulates at base of corolla.

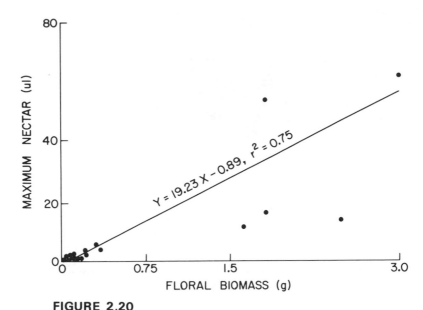

FIGURE 2.20

Linear regression of medium to large bee-pollinated plants: floral biomass (g) vs. maximum nectar (μl). Each point represents the values for a single species (see also table 2.7).

open pumice savanna and evergreen riparian forest. Light availability is almost never limiting in the former, but is often limiting in the latter. Soil moisture is seldom limiting in the latter, but is extremely low during the dry season in the former. Soil depths are greatest in the evergreen riparian forest but extremely shallow in open pumice savanna.

As a consequence, flowering in open pumice savanna occurs primarily in the wet season, although some trees there flower in the dry season. Flowering in riparian forest is more or less equitably dispersed throughout the year. The herbaceous life form is almost absent from such forest, presumably due to year-round low light availability near ground level.

Other habitats present differing combinations of the above two extremes. Hillside dry forest presents low soil moisture, but high light availability in the dry season, and the inverse during the wet sea-

TABLE 2.7
NECTAR PRODUCTION AND FLORAL FEATURES OF
NECTARIFEROUS PLANTS ADAPTED PRINCIPALLY FOR
MEDIUM TO LARGE BEE POLLINATION

Species	Family	Floral Biomass (g)	Corolla Length (cm)	Maximum Nectar (μl)
Stemmadenia obovata	Apocynaceae	3.0	7.2	66.0
Thevetia ovata	Apocynaceae	1.8	7.2	58.0
Styrax argenteus	Styracaceae	0.25	2.0	13.7
Centrosema plumieri	Fabaceae	1.9	3.8	13.0
Tabebuia rosea	Bignoniaceae	1.6	5.8	11.5
Gliricidia sepium	Fabaceae	0.3	2.3	5.3
Caesalpinia eriostachys	Caesalpiniaceae	0.36	1.9	3.6
Tabebuia neochrysantha	Bignoniaceae	0.20	5.2	3.4
Centrosema pubescens	Fabaceae	0.20	2.5	2.5
Petastoma patelliferum	Bignoniaceae	0.10	3.0	2.5
Andira inermis	Fabaceae	0.34	1.0	2.0
Piscidia cartheguensis	Fabaceae	0.18	1.8	0.8
Pterocarpus rohrii	Fabaceae	0.08	1.5	0.6
Dalbergia retusa	Fabaceae	0.07	1.4	0.6
Phaseolus atropurpureus	Fabaceae	0.10	2.3	0.5
Myrospermum frutescens	Fabaceae	0.10	1.7	0.4
Stachytarpheta frantzii	Verbenaceae	0.75	1.7	0.4
Ipomaea nil	Convolvulaceae	0.15	5.0	0.3
Hyptis suaveolens	Labiatae	0.01	0.7	0.2
MEAN ($\bar{X}$)		0.60	3.05	9.75

Nectar production and floral features of nectariferous plants adapted principally for medium to large bee pollination.

son. There are many trees and some lianas-not limited by low surface
soil moisture that flower during the dry season, but most plants representing other life forms flower early in the wet season, a time of
ready moisture availability and adequate light availability before the
canopy has fully leafed out. Open flat savanna, seasonal swamps,
pastures, and roadside edge are very similar to pumice savanna in
their soil and light availability patterns, and, as a consequence, are
composed largely of nectariferous herbs which flower in the latter
half of the wet season, e.g., *Turnera* spp., *Heliotropium*, and *Krameria*.

Differences in nectar availability between sites of the same habitat

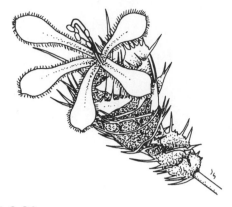

FIGURE 2.21
Loeselia ciliata (Polemoniaceae), a butterfly-pollinated herb. Nectar accumulates at base of corolla.

FIGURE 2.22
Asclepias curassavica (Asclepiadaceae), a butterfly-pollinated herb. Nectar accumulates inside corolla in 'nectar cups'.

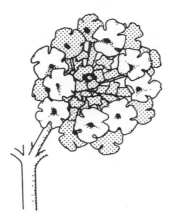

FIGURE 2.23
Lantana camara (Verbenaceae), a butterfly-pollinated shrub. Nectar accumulates inside corolla tube. Yellow flowers are functional nectariferous flowers of the day, while orange and red flowers are those of previous days, devoid of nectar, but still serving to augment pollinator
attraction.

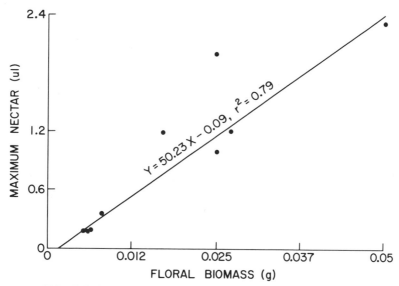

FIGURE 2.24
Linear regression of butterfly-pollinated plants: floral biomass (g) vs.
maximum nectar (μl). Each point represents the values for a single species (see also table 2.8).

TABLE 2.8
NECTAR PRODUCTION AND FLORAL FEATURES OF
NECTARIFEROUS PLANTS ADAPTED PRINCIPALLY FOR
BUTTERFLY AND SETTLING MOTH POLLINATION

Species	Family	Floral Biomass (g)	Corolla Length (cm)	Maximum Nectar (μl)
Butterfly Flowers				
Cordia pringlei	Boraginaceae	0.05	1.9	2.3
Justicea sp.	Acanthaceae	0.025	2.6	2.0
Asclepias curassavica	Asclepiadaceae	0.028	0.6	1.2
Calycophyllum candidissimum	Rubiaceae	0.025	0.5	1.0
Lantana camara	Verbenaceae	0.008	1.0	0.4
Loesilia ciliata	Polemoniaceae	0.005	1.6	0.2
Blechum costaricensis	Acanthaceae	0.006	1.6	0.2
Cordia dentata	Boraginaceae	0.015	0.9	0.1
MEAN ($\overline{X}$)		0.02	1.33	0.93
Settling Moth Flowers				
Cedrela odorata	Meliaceae	–	0.92	1.70
Guarea excelsum	Meliaceae	–	0.65	1.40
Cordia alliodora	Boraginaceae	0.03	1.3	0.94
Pithecellobium longifolium	Mimosaceae	–	2.3	0.81
Enterolobium cyclocarpum	Mimosaceae	0.003	1.1	0.18
Anacardium excelsum	Anacardiaceae	0.025	0.6	0.03
MEAN ($\overline{X}$)		0.019	1.145	0.84

Nectar production and floral features of nectariferous plants adapted principally for butterfly and settling moth pollination.

type are often extreme due to local floristic differences and the extreme patchiness of precipitation. Some appreciation of differences in forest composition within Guanacaste may be gained by reference to Holdridge et al. (1971), while each of my study areas at Comelco Ranch displayed widely varying floristic composition (Frankie et al. 1974). As an example, at different sites along the Rio Corobici near Cañas, riparian forest understory is dominated by *Quassia amara* in several instances, *Psychotria cartheginensis* in another, by *Randia spinosa* in another, and by *Allophyllis occidentalis* in still another.

FIGURE 2.25
Enterolobium cyclocarpum (Mimosaceae), a settling moth-pollinated tree. Nectar accumulates inside corolla tubes.

Similarly, different trees may dominate the overstory of each dry hillside forest site, and so on.

Influences of local patchiness in precipitation on flowering seasonality, and thereby nectar availability, have been described by Opler et al. (1976) for two sites in Guanacaste. The onset of flowering by many shrubs, lianas, and trees at the end of the dry season may differ by a month or more locally, dependent upon the occurrence of small token precipitation events. Other species with early dry season flowering periods depend upon token rainfall episodes during the first two dry season months (December and January). For two years such episodes did not occur on portions of our study area and some plants, e.g., *Tabebuia palmeri*, did not flower at all.

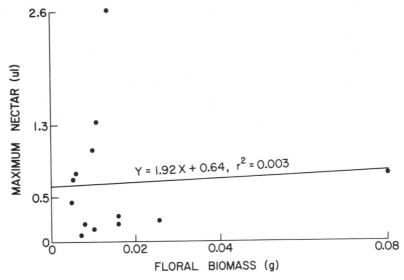

FIGURE 2.26
Linear regression of small bee and wasp-pollinated plants: floral biomass (g) vs. maximum nectar (μl). Each point represents the values for a single species. (see also table 2.9).

Between year differences in nectar availability patterns in Guanacaste are also significant, largely due to differences of amount and timing of precipitation. In addition, one common tree, *Andira inermis*, flowers synchronously in the dry season of even-numbered years. Since this tree is an important nectar source for many bees (Frankie et al. 1976), the virtual absence of its flowers during odd-numbered years is a significant factor in the province's yearly nectar supply.

Plants with flowers adapted to the different pollinator groups have quite different patterns of seasonality (fig. 2.28) (see also Frankie 1975). Of course, each seasonal pattern is comprised of the combined flowerings of all nectariferous plants adapted to each group, and thus, does not represent the flowering of individual plants or the availability patterns in a specific habitat (see above).

Most bat plants flower in the dry season (January to May) with the peak consisting of a sequence of flowerings by different species (Heithaus et al. 1975).

TABLE 2.9
NECTAR PRODUCTION AND FLORAL FEATURES OF
NECTARIFEROUS PLANTS ADAPTED PRINCIPALLY FOR
SMALL BEE AND WASP POLLINATION

Species	Family	Floral Biomass (g)	Corolla Length (cm)	Maximum Nectar (µl)
Forsteronia spicata	Asclepiadaceae	0.015	0.5	2.6
Cordia collococca	Boraginaceae	0.011	0.4	1.32
Trichilia cuneata (Staminate)	Meliaceae	0.010	0.3	1.02
Euphorbia heterophylla	Euphorbiaceae	0.08	0.5	0.75
Roupala complicata	Proteaceae	0.006	0.8	0.75
Cissus sicyoides	Vitaceae	0.005	0.2	0.70
Cissus rhombifolia	Vitaceae	0.005	0.2	0.44
Simarouba glauca	Simaroubaceae	0.015	0.5	0.31
Casearia aculeata	Flacourtiaceae	0.025	0.5	0.25
Casearia nitida	Flacourtiaceae	0.015	0.3	0.20
Bursera tomentosa (Staminate)	Burseraceae	0.008	0.3	0.19
Cordia panamensis	Boraginaceae	0.010	0.5	0.12
Cordia inermis	Boraginaceae	–	0.3	0.08
Waltheria indica	Sterculiaceae	0.007	0.5	0.06
MEAN ($\overline{X}$)		0.016	0.41	0.63

Nectar production and floral features of nectariferous plants adapted principally for small bee and wasp pollination.

There appears to be no distinct peak for hummingbird plants, there
being an equitable distribution of seven to ten species in flower each
month. Some of these plants have distinctly seasonal flowerings, e.g.,
Combretum farinosum (Combretaceae), while others flower spo-
radically throughout the year, e.g., *Malvaviscus arboreus* (fig. 2.12).
By contrast, in the Atlantic wet forest there is a distinct trough of hummingbird plant flowering from October to December (Stiles 1978).

An intense investigation of hawkmoth adapted plants in Guanacaste is currently in progress (Haber and Frankie, in preparation),
and no attempt will be made to summarize the seasonality of those
plants at present.

For medium and large bee adapted plants there is a distinct abundance of nectar resources from February through July, encompassing

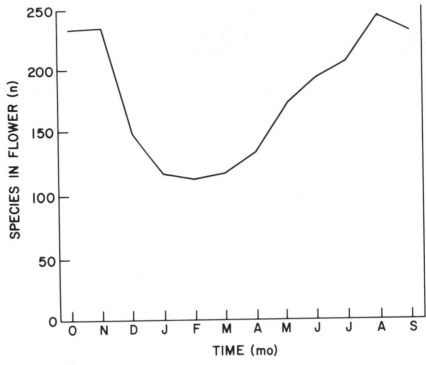

FIGURE 2.27
Seasonal phenology of nectariferous plants adapted to any pollinatory group. Pattern is strongly influenced by the large number of herbaceous plants.

the latter half of the dry season and the first half of the wet season (Frankie 1976, Frankie et al. 1982).

The wet season is the period when most butterfly-adapted plants are in flower. By way of contrast, settling moth adapted plants seem to have two indistinct peaks of flowering, one during January and February (mid dry season) and another in July (early wet season).

Plants adapted to wasps and small bees have their peak flowering during early dry season (November to January), a time when plants adapted to no other pollinatory group are in peak flower (Frankie 1975).

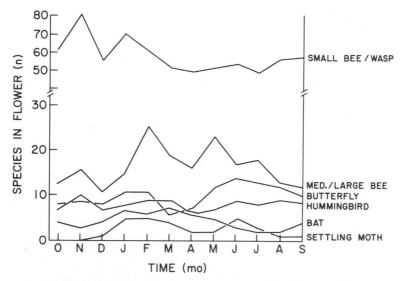

FIGURE 2.28
Seasonal flowering phenology of nectariferous plants by pollinatory syndrome. Hawkmoth plants excluded due to study in progress.

THE NECTAR-FEEDING ANIMALS

The 1,289 nectarivorous animals encountered during my study are listed by class, order, and family in table 2.10 (this figure is approximate because determinations of all species are not complete). It should be noted that two large mammals, the four-eyed opossum (*Metachirops opossum*) and the kinkajou (*Potos flavus*), were observed feeding on nectar, but were judged as only sporadic nectarivores.

For purposes of the present account, I will usually refer to nectarivores in broad groupings as follows: bats, hummingbirds, hawkmoths, medium to large bees, butterflies, nocturnal settling moths, and small bees, including wasps. The remaining diurnal settling moths, beetles, bugs, and flies are included in category small bees above, and this category may be referred to as one of small to medium generalized flower visitors.

Each nectarivore group has its own unique combination of body size, foraging technique, foraging time, and life cycle features that

TABLE 2.10
FLORAL NECTAR-FEEDING ANIMALS OF GUANACASTE PROVINCE, COSTA RICA

Class — Order — Family	Species (N)	Class — Order — Family	Species (N)
Mammalia		Heteroptera	
Chiroptera		Largidae	1
Phyllostomatidae	10	Lygaeidae	3
Aves		Pyrrhocoridae	3
Apodiformes		Hymenoptera	
Trochilidae	8	Anthophoridae	±60
Passeriformes		Apidae	36
Parulidae	1	Braconidae	±3
Icteridae	2	Chalcidae	±10
Insecta		Chrysididae	±3
Lepidoptera		Formicidae	±5
Arctiidae	2	Halictidae	40
Ctenuchidae	10	Megachilidae	±40
Glyphipterygidae	1	Mutillidae	±2
Hesperiidae	±200	Pompilidae	37
Lycaenidae	115	Scoliidae	±5
Noctuidae	±50	Sphecidae	±75
Nymphalidae	35	Tiphiidae	±6
Papilionidae	14	Vespidae	±40
Pieridae	23	Colletidae	1
Pyralidae	±25	Diptera	
Riodinidae	18	Bombyliidae	42
Satyridae	2	Conopidae	15
Scythridae	1	Culicidae	±5
Sesiidae	3	Mydaidae	1
Sphingidae	±100	Nemestrinidae	1
Stenomidae	1	Stratiomyidae	15
Yponomeutidae	1	Tabanidae	1
Coleoptera		Tachinidae	±100
Cerambycidae	31	Tipulidae	±5
Brenthidae	2	Syrphidae	55
Bruchidae	±5	TOTAL	±1292
Chrysomelidae	±15		
Mordellidae	3		
Rhipiphoridae	4		

Representation of nectarivorous animals in Guanacaste Province, Costa Rica, listed by class, order, and family.

have resulted in the correlative evolutionary adaptation by plants. Through evolutionary time, each plant's floral features, including nectar production and chemistry, have formed an adaptive complex, best suited to the plant's needs for optimal fertilization by attracting one or more flower visitor groups (Baker and Hurd 1968; Proctor and Yeo 1973).

The following is a brief introduction to the primary nectarivore categories found in Guanacaste in order by decreasing size. Ten species of phyllostomatid bats are nectarivores in Guanacaste, although only six may be viewed as regular flower visitors (Heithaus et al. 1975). Important attributes of bats as flower visitors are their nocturnal activity, ability to forage singly or in groups (Heithaus et al. 1974; Baker 1972), and to feed on alternate resources, i.e., fruits or insects, when floral nectar is unavailable.

Eight hummingbirds, including the migratory ruby-throated hummingbird (*Archilochus colibri*), are regular nectarivores in Guanacaste Province. About half of these are normally territorial and the remainder seem to fit the role of trap-liners (Stiles and Wolf 1970; Feinsinger 1976; Stiles 1975). Hummingbirds are notable as being the largest bodied regular diurnal nectarivores.[5]

Currently, a detailed study of the approximately 100 hawkmoth species is being carried out in Guanacaste by Haber and Frankie (unpublished). Aspects of foraging behavior, seasonality, and life history information will be integrated with data on their nectar sources. Little can be stated at this time, except for the fact that most hawkmoths are nocturnal, with only three or four relatively small species foraging in daylight hours.

The large bee category (those species with body lengths greater than 1.2 cm) comprises approximately 40 species representing three families, Apidae-*Bombus* (2), *Euglossa* (3), *Eulaema* (4), *Euplusia* (2), *Exarete* (1); Anthophoridae-*Centris s. l.* (20), *Xylocopa* (8); and Colletidae-*Ptiloglossa* (1). These bees may be either opportunistic or trap-lining foragers and may also fly long distances (Janzen 1971;

---

5. Several species of orioles (*Icterus* spp.) forage on flowers of *Erythrina glauca, Tabebuia neochrysantha*, and a few others. They are larger bodied than the hummingbirds, but are only incidental members of the flower visitor community.

Heinrich 1976; Frankie 1976; Frankie et al. 1976). *Ptiloglossa* is a nocturnal or crepuscular forager, the remainder being diurnal.

When considering the flower-visiting habits of bees, it is important to remember that, while both sexes feed on flower nectar, females may visit some flowers only to gather pollen for their brood.

Medium-sized bees (those species with body lengths between 0.8 and 1.2 cm) visit the same flowers that are visited by large bees, more often than they visit small bee/wasp flowers. Approximately 50 species of Anthophoridae, Apidae, Halictidae, and Megachilidae fall into this group; thus the entire medium-large bee category is composed of about 90 species.

Approximately 400 species of butterflies representing seven families regularly feed on floral nectar in Guanacaste.[6] Butterflies are characterized by their long proboscuses and diurnal habits, although a few skippers (Hesperiidae) feed just at dawn or sunset. Most butterflies feed on nectar between 8.00 and 15.00 hours, beginning later and stopping earlier than other diurnal nectarivores. Some butterflies pass the dry season as nonreproductive adults, whle others emigrate to adjacent higher, moister elevations to pass this unfavorable season.

The settling moth category is poorly understood in Guanacaste, being comprised of perhaps 300 nocturnal moths belonging primarily to the Noctuidae and Pyralidae. They range in size from the huge *Erebus odorus* (Noctuidae) to tiny species of Nymphulinae (Pyralidae).

The wasp/small bee category is composed of the smallest sized pollinators, although a few wasps, e.g., *Polistes* (Vespidae) and *Pepsis* (Pompilidae) are quite large. Perhaps as many as 300 species of small bees and wasps are included here, including about a dozen *Trigona* (Apidae), probably the dominant nectarivore group in the province in terms of both numbers and biomass. *Trigona* are normally found in or adjacent to forested habitats. They are absent or rare in deforested habitats.

Also included in the wasp/small bee category, which could just as appropriately be termed the small generalized nectarivore category,

6. The adults of all Brassolidae, most Satyridae, many Nymphalidae, and some representatives of other butterfly families (except Papilionidae and Pieridae) never feed on floral nectar. Instead they feed as adults on rotting fruit, fungi, bird droppings, excrement, and sap exudates. Adults of some *Heliconius* (Nymphalidae) feed on pollen as well as nectar.

are in the neighborhood of 250 nectarivorous Diptera, Coleoptera, and diurnal Lepidoptera (exclusive of butterflies).

PHYSICAL CHARACTERISTICS OF NECTARIVORES

When one considers physical characteristics of flowers, such as nectar volume, floral biomass, and corolla length, one may ask: "How do these features compare with those of the flower visitors?" In this section I consider some of those visitor features which may be related in turn to floral characteristics.

Some physical characteristics of the Guanacaste flower visitor groups are summarized in table 2.11. Of these visitors, biomass (g) will be the most obvious correlate of maximum nectar and floral biomass in later discussion,[7] while body length (cm) and mouthpart length (cm) may be best related to corolla length.

Bats have by far the greatest biomass of any visitor group, averaging more than 32 g. Among the other groups, only hummingbirds (4.04g) and hawkmoths (1.05g) weighed more than a gram, while the other insect visitor groups each averaged less than a half gram (table 2.11).

Data on body length (cm) was available for a larger number of insect visitors, but was not available for the vertebrates (table 2.11). However, body lengths for the vertebrates were estimated from field guides, field experience, and other considerations and are presented together with those for insects (fig. 2.28). As can be readily seen, the distribution of visitor body length is log-normal, with a modal value between 0.5 and 0.6 cm-in the size range of small bees, small butterflies, and small settling moths.

The order of body lengths is exactly the same as the order of visitor biomasses for the various visitor groups, but the biomass:body length ratios vary considerably (table 2.11). For example, the mean body lengths of large bees (1.55 cm) and butterflies (1.54 cm) are nearly identical, yet their biomass:length ratios are 0.29 g/cm and 0.12 g/cm, respectively. This is an obvious demonstration of the fact that bees

7. Weights for bats were obtained from Goodwin and Greenhall (1961), while those for hummingbirds were extracted from Feinsinger (1976). Biomass values for the insectan visitor groups were obtained by weighing freshly killed individuals with a Pesola 50 g balance. Numerous individuals of small species were weighed together to obtain a more accurate weight.

TABLE 2.11
SOME PHYSICAL CHARACTERISTICS OF NECTARIVOROUS ANIMAL GROUPS ADAPTED TO PLANTS WITH DIFFERENT POLLINATORY SYNDROMES

Pollinator Group	Visitor Biomass (g)			Body Length (cm)			Mouthpart Length (cm)			Visitor Biomass		Nectar Visitor Biomass
	X	S.D.	N	X	S.D.	N	X	S.D.	N	Body Length	Mouthpart Length	Biomass
Bats	32.35	24.53	8	–	–	–	–	–	–	–	–	40.51
Hummingbirds	4.05	1.69	7	–	–	–	2.54	0.76	7	–	1.59	4.16
Hawkmoths	1.05	0.07	2	4.35	0.10	2	5.31	2.67	2	0.24	0.20	124.33
Large bees	0.45	0.27	4	1.55	0.41	44	0.56	0.44	28	0.29	0.80	21.67
Butterfly	0.18	0.10	38	1.54	0.68	159	1.31	0.50	34	0.12	0.14	5.17
Wasp/Small bees	0.10	0.15	17	0.86	0.47	62	0.24	0.26	19	0.12	0.42	6.3
All visitors	3.94	12.41	76	–	–	–	1.02[a]	1.06	90	–	0.59[a]	46.79

Some physical characteristics of nectarivorous animal groups adapted to plants with different pollinatory syndromes.

[a] Excludes bats

have relatively short robust bodies, while those of butterflies are relatively long and slender.

Mouthpart lengths are not positively correlated with visitor biomass or body length (table 2.11).[8] On average, mouthpart length averages 1.07 cm for all flower visitors, however, the real value for the community will be lower due to the relatively small proportion of small bees and wasps included in our sample. Naturally, when one considers the relatively large numbers of individuals of small bodied species in the community, this value would be skewed even lower. Hawkmoths had the greatest average mouthpart length (5.31 cm), while hummingbirds (2.54 cm), butterflies (1.31 cm), and large bees (0.56 cm) are usually thought of as "long-tongued" visitors. Wasps and small bees (0.24 cm) had the shortest mean mouthpart length.

If one examines the ratio between biomass and mouthpart length (g/cm) in table 2.11, one finds, on average, there are 0.59 g biomass per cm mouthpart length. Hummingbirds had the greatest biomass: mouthpart length ratio (1.59 g/cm), while butterflies, with their long probosces, had the lowest (0.14 g/cm).

Another important characteristic of each flower visitor group, vis-à-vis their utilization of nectar, is their means of thermoregulation. Bats, hummingbirds, hawkmoths, and some large bees regulate their body temperatures by physiological means internally, while the remaining visitor groups regulate their temperatures behaviorally by more passive means.

Longevity is another crucial feature of various visitor taxa as relates to nectar utilization. Nectarivorous bats and birds normally have life expectancies exceeding a year, and must have a continual supply of sustenance. To accomplish this when local nectar supplies are inadequate, they may either switch their attentions to alternate resources or emigrate. Bats and birds do both. Heithaus et al. (1975) have shown that bats feed on nectar in the Guanacaste dry season, but switch to frugivory in the wet season. *Phyllostomus discolor* apparently emigrates in the wet season when appropriate nectar sources are absent, while *Phyllostomus hastatus*, the largest nectarivore in the commun-

---

8. Mouthpart lengths were not available for bats. By observation some, such as that of *Glossophaga soricina*, may exceed the shortest of hawkmoth probosces.

ity, is omnivorous, feeding on other bats, insects and fruits, in addition to floral nectar.

Hummingbirds, although they occasionally feed on insects as an alternate resource, more often emigrate to other portions of Costa Rica during some portions of the year. *Archilochus colibri*, which breeds in North America, is a dry season migrant to Guanacaste (as well as to other portions of Central America).

The insect nectarivores all have maximum adult longevities of less than a year. In fact, most survive for periods of less than a month, while only some butterflies, and possibly some large bees and hawkmoths may survive for periods exceeding a month. Very few species may have maximum longevities in the range of four to six months. Such life spans have been documented only for butterflies (for a review refer to Scott 1974).

As a consequence of their short life spans, most insect nectarivores have their flight periods adapted to optimal yearly periods of nectar availability, as modified by other considerations such as availability of larval food resources and interspecific competition.

FLOWER-ANIMAL INTERRELATIONSHIPS

Although there are many flower–visitor interactions one might broach with regard to the Guanacaste ecosystem, I will restrict myself to those relating to nectar utilization. More specifically, the relationships between floral features and animal characteristics will receive my greatest scrutiny.

There is a clear, positive relationship between nectar production and pollinator size. Earlier, I demonstrated the close relationship between flower size, as either biomass or corolla length, and nectar production (tables 2.2–2.9; figs. 2.7, 2.13, 2.15, 2.20, 2.24, 2.26). If one combines the regression lines (fig. 2.1), it can be immediately seen that the plants adapted to particular visitor groups are arranged along the overall regression (table 2.3) in order of visitor biomass, with the exception of hawkmoth plants which provide more nectar per flower, on average, than do the flowers adapted for the heavier-bodied hummingbirds. Bat plants occur at the upper right of the regression, and small bee/wasp plants at the lower left.

Another view of the relationship between nectar production and nectarivores is gained by comparing the frequency distributions for corolla length—generally a positive function of nectar amount—and visitor body lengths (tables 2.2 and 2.11, figs. 2.29–2.30). The range of the above variables is nearly identical, and the type of distribution

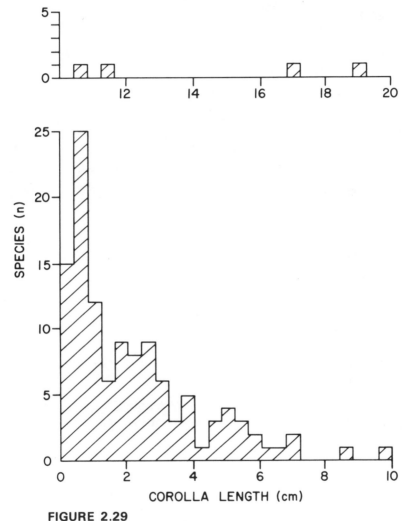

FIGURE 2.29
Frequency distribution for corolla lengths of 121 selected nectariferous plants in Guanacaste Province, Costa Rica. Note log-normal distribution.

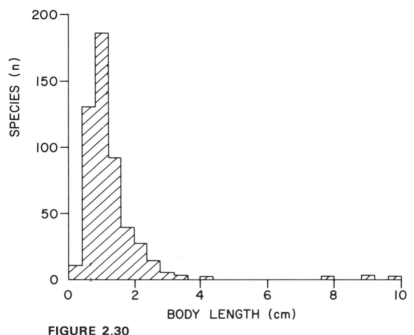

FIGURE 2.30

Frequency distribution for body lengths of 514 nectarivorous animals in Guanacaste Province. Note log-normal distribution.

(log-normal) is similar for both. It may be noted, however, that the most frequent corolla length is smaller than that for the animal visitors, while in opposition corolla lengths are generally larger (figs. 2.29–2.30).

Further insight into the relation between floral nectar production and visitors may be gained by examining the ratio between maximum nectar and visitor biomass, as either $\mu l/g$ or $g/\mu l$ (tables 2.2, 2.11). On average, for the plants examined there were almost 47 μl maximum nectar per flower per gram of visitor weight. The values for the different visitor groups ranged from 4.16 $\mu l/g$ for hummingbird flowers to 124.33 $\mu l/g$ for hawkmoth flowers.

If one corrects the nectar per visitor biomass figures by multiplying nectar volume by mean sugar concentration, as has been done elsewhere in this work (see paper 3), one would find the variability considerably reduced, since bat and hawkmoth flowers usually have low sugar concentrations, while bee, wasp, and butterfly flowers usually have higher sugar concentrations (see paper 4). Even still, the amount of sugar available per gram body weight would be higher for bats, hawkmoths, and medium to large bees (see paper 3).

The larger amount of nectar (or sugar) per gram body weight in flowers adapted for bats, hawkmoths, and medium to large bees is not surprising, for it is just these animals which are both the long-distance foragers and physiological thermoregulators among the Guanacaste flower visitor communty (Heithaus et al. 1975; Haber, unpub.; Janzen 1971; Frankie et al. 1976). Hummingbirds, although physiological thermoregulators, are usually territorial in their foraging behavior within this ecosystem (Stiles and Wolf 1970; Opler, unpub.).

ACKNOWLEDGMENTS

All of the results reported here were part of the O.T.S. Comparative Ecosystem Study funded by the National Science Foundation (Grants GB-25592, GB-25592A No. 2 and BMS 73-01619 A01; H. G. Baker and G. W. Frankie, principal investigators), and the assistance of both the N.S.F. and the Organization for Tropical Studies is gratefully acknowledged.

No major study in recent years can be attributed to a single scientist alone, and I must therefore point out the major roles played by H. G. Baker, I. Baker, K. S. Bawa, and G. W. Frankie. Other biologists in Costa Rica during the study contributed to ideas and techniques expressed herein, particularly B. Bentley, R. Carroll, P. Feinsinger, E. R. Heithaus, S. P. Hubbell, and F. G. Stiles.

A debt of gratitude is due the many botanists (together with their institutions) who provided plant identifications, as well as those who provided logistical assistance.

REFERENCES

Baker, H. G. 1972. Evolutionary relationships between flowering plants and animals in American and African tropical forests. In B. J. Meggers, E. S.

Ayensu, and W. D. Duckworth, eds., *Tropical Forest Ecosystems in Africa and South America*, pp. 145–159. Washington, D.C.: Smithsonian Institution Press.

Baker, H. G. and P. D. Hurd, Jr. 1968. Intrafloral ecology. *Ann. Rev. Entom.* 13:385–414.

Bawa, K. S. and P. A. Opler. 1975. Dioecism in tropical forest trees. *Evolution* 29:167–179.

Cronquist, A. J. 1968. *The Evolution and Classification of Flowering Plants.* Boston: Houghton-Mifflin.

Faegri, K. and L. Van der Pijl. 1966. *The Principles of Pollination Biology.* Oxford: Pergamon Press.

Feinsinger, P. 1976. Organization of a tropical guild of nectarivorous birds. *Ecol. Monogr.* 46:257–291.

Frankie, G. W. 1975. Tropical forest phenology and pollinator plant coevolution. In L. E. Gilbert and P. H. Raven, eds., *Coevolution of Animals and Plants*, pp. 192–209. Austin: University of Texas Press.

Frankie, G. W. 1976. Pollination of widely dispersed trees by animals in Central America, with an emphasis on bee pollination systems. In J. Burley and B. T. Styles, eds., *Variation, Breeding, and Conservation of Tropical Forest Trees*, pp. 151–159. London: Academic Press.

Frankie, G. W., H. G. Baker, and P. A. Opler. 1974. Comparative phenological studies of trees in tropical wet and dry forests in the lowlands of Costa Rica. *J. Ecol.* 62:881–919.

Frankie, G. W., P. A. Opler, and K. S. Bawa. 1976. Foraging behavior of solitary bees: Implications for outcrossing of a neotropical forest tree species. *J. Ecol.* 64:1049–1057.

Frankie, G. W., W. A. Haber, P. A. Opler, and K. S. Bawa. 1982. Characteristics and organization of the large bee pollination system in the Costa Rican dry forest. In C. E. Jones and R. J. Little, eds., *Handbook of Experimental Pollination Biology*. New York: Van Nostrand.

Gentry, A. H. 1974. Coevolutionary patterns in Central American Bignoniaceae. *Ann. Missouri Bot. Gard.* 61:728–759.

Gentry, A. H. 1976. Bignoniaceae of southern Central America: Distribution and ecological specificity. *Biotropica* 8:117–131.

Goodwin, G. G. and A. M. Greenhall. 1961. A review of the bats of Trinidad and Tobago. *Bull. Amer. Mus. Nat. Hist.* 122:187–302.

Heinrich, B. 1975. The role of energetics in bumblebee-flower interactions. In L. E. Gilbert and P. H. Raven, eds., *Coevolution of Animals and Plants*, pp. 141–158. Austin: University of Texas Press.

Heinrich, B. 1976. The foraging specializations of individual bumblebees. *Ecol. Monogr.* 46:105–128.

Heithaus, E. R., T. H. Fleming, and P. A. Opler. 1975. Foraging patterns and resource utilization in seven species of bats in a seasonal tropical forest. *Ecology* 56:841–854.

Heithaus, E. R., P. A. Opler, and H. G. Baker. 1974. Bat activity and pollination of *Bauhinia pauletia:* Plant-pollinator coevolution. *Ecology* 55:412–419.

Holdridge, L. R. 1967. *Life Zone Ecology*. rev. ed. San Jose, Costa Rica: Tropical Science Center.

Holdridge, L. R., W. C. Grenke, W. H. Hatheway, T. Liang, and J. A. Tosi, Jr. 1971. *Forest Environments in Tropical Life Zones:* A Pilot Study. Oxford: Pergamon Press.

Janzen, D. H. 1971. Euglossine bees as long-distance pollinators of tropical plants. *Science* 171:203–205.

Janzen, D. H. 1973. Sweep samples of tropical foliage insects: Effects of seasons, vegetation types, time of day and insularity. *Ecology* 54:687–708.

Opler, P. A., H. G. Baker, and G. W. Frankie. 1975. Reproductive biology of some Costa Rican *Cordia* species. (Boraginaceae). *Biotropica* 7:234–247.

Opler, P. A., H. G. Baker, and G. W. Frankie. 1977. Recovery of tropical lowland ecosystems. J. Cairns, Jr., K. L. Dickson, and E. E. Herricks, eds. In *Recovery and Restoration of Damaged Ecosystems*, pp. 379–419. Charlottesville, University Press of Virginia.

Opler, P. A., G. W. Frankie, and H. G. Baker. 1976. Rainfall as a factor in the release, timing, and synchronization of anthesis by tropical trees and shrubs. *J. Biogeography* 3:231–236.

Percival, M. S. 1965. *Floral Biology*. Oxford: Pergamon Press.

Proctor, M. and P. Yeo. 1973. *The Pollination of Flowers*. London: Collins.

Scott, J. A. 1974. Lifespan of butterflies. *J. Res. Lepid.* 12:225–230.

Stiles, F. G. 1975. Ecology, flowering phenology, and pollination of some Costa Rican *Heliconia* species. *Ecology* 56:285–301.

Stiles, F. G. 1978. Temporal organization of flowering among the hummingbird foodplants of a tropical wet forest. *Biotropica* 10:194–210.

Stiles, F. G. and L. L. Wolf. 1970. Hummingbird territoriality at a tropical flowering tree. *Auk* 87:465–492.

Turner, D. 1975. *The Vampire Bat*. Baltimore: Johns Hopkins University Press.

3

PATTERNS OF NECTAR PRODUCTION AND PLANT-POLLINATOR COEVOLUTION

ROBERT WILLIAM CRUDEN
UNIVERSITY OF IOWA
SHARON MARIE HERMANN
UNIVERSITY OF IOWA; UNIVERSITY OF ILLINOIS
STEVEN PETERSON
UNIVERSITY OF COLORADO

There is now general agreement that floral nectars play an important role in plant pollinator interactions, but until the past decade our understanding of floral nectars was, with hindsight, simplistic; nectars contained sugar that pollinators utilized and the energetic reward reflected the energetic requirements of the pollinators. This uncomplicated view of nectar and its function was shattered by the demonstration that nectars contained significant amounts of nutrients other than sugars, for example, amino acids and lipids, and that the kinds and amounts of these nutrients reflected coevolution with the plant's pollinators (Baker and Baker 1973a, b, 1975, 1976; Cruden and Hermann-Parker 1979; Cruden and Toledo 1977). Although the constituents of floral nectars are now being studied, other facets of nectar and its production remain neglected.

With respect to the amount of nectar produced per flower, numerous workers have observed that flowers pollinated by high-energy requiring animals, for example, bats, hawkmoths and birds, produce significantly more nectar than flowers pollinated by low-energy requiring animals, such as butterflies, bees, and flies. Although most pollination biologists accept this broad generalization, there is little quantitative data to support it (but see Beutler 1930; Heinrich 1975a, c, 1976a, b; Hocking 1968; Kleber 1935 for amounts of nectar in some bee flowers, and Fahn 1949; Opler 1982; and Percival 1965 for other flower classes). In addition, the amount of nectar produced per flower may be subject to selective pressures other than pollinator class. For example, flower density (Heinrich and Raven 1972), habitat, and breeding systems (Cruden 1976a, b), and nectar thieves can influence nectar production. These have, in general, received little attention.

In contrast to the constituents of floral nectars and their adaptive significance, which have recently received considerable attention (e.g., Baker 1975, 1978; Baker and Baker 1973a, b, 1975, 1976, 1982; Hainsworth and Wolf 1976; Percival 1962; Stiles 1976), adaptive patterns of nectar production have received minimal and sporadic attention from pollination biologists (Andrejeff 1932; Bawa and Opler 1975; Beutler 1930; Carpenter 1976; Cruden 1976a, b; Frankie et al. 1976; Heinrich 1975b, c; Hocking 1968; Kleber 1935; Schaffer and Schaffer 1977). It is generally accepted that flowers pollinated by diurnally active animals produce nectar during the day and that flowers pollinated by nocturnally active animals produce nectar at night. Other adaptive responses have been reported, including resumption of nectar production following nectar removal (Raw 1953; Wykes 1950); cessation of nectar secretion when pollinators are inactive (Bonnier 1878); cessation of secretion following pollination (Bonnier 1878; Pankratova 1950); and resorption of nectar in old or pollinated flowers (Boëtius 1948; Bonnier 1878; Pankratova 1950).

The major objective of this paper is to outline in broad terms what we perceive to be the basic pattern of nectar production in a typical flower. We hope this report will stimulate research, provide others with testable hypotheses, and create a framework within which future work can be interpreted and compared. In particular we will: 1) provide evidence that the amount of nectar produced per flower reflects the energetic demands of the plant's pollinators and discuss other factors that affect the amount and quality of nectar produced by a flower; 2) propose a model for nectar secretion, discuss exceptions to the model, and indicate the adaptive nature of those exceptions; 3) discuss variation in rates of nectar secretion and how these are adaptive; and 4) show that the pattern of nectar secretion is integrated with other facets of a plant's reproductive biology. The influence of various environmental parameters on nectar secretion is briefly discussed.

MATERIALS AND METHODS

Nectar measurements were made in the field, primarily in the southwestern United States and Mexico. We usually selected for study species whose flowers have nectaries at the base of a tubular corolla, because nectar in open nectaries is subject to evaporation or dilution by rain (Andrejeff 1932; Beutler 1930; Boëtius 1948; Fahn 1949; Kleber 1935; Park 1929; Scullen 1940). With the exception of *Spathodea campanulata* and the sunbird-pollinated species, all the species were studied in their native habitats.

Time of arrival of pollinators was selected as a point of comparison because, in most instances, it is a period of similar activity. The pollinators have just resumed activity following a long period of inactivity and they immediately forage for food. Thus, we reasoned, available nectar should be maximal. By choosing this time we simplified the logistics. First, we did not have to bag flowers as most of the species we studied were rarely visited by species other than their normal pollinators. Second, our sampling was not restricted to a limited number of bagged individuals. When sampling was conducted after the pollinators became active, mosquito netting was used to exclude flower visitors.

Nectar samples were collected to provide an accurate estimate of the nectar available in a population of flowers when pollinators first visited the flowers: 1) We sampled flowers in all stages of development because the amount of nectar may vary with stage of development (Fahn 1949; Pankratova 1950) and between sexes of monoecious and dioecious species (Bawa and Opler 1975; Fahn 1949). Differences between staminate and pistillate phases in a number of protandrous species were not significant (table 3.1) and the data in appendix I in-

TABLE 3.1
AMOUNTS OF SUGAR IN THE NECTAR OF DICHOGAMOUS
FLOWERS AT THE START OF POLLINATOR ACTIVITY

		Staminate Phase		Pistillate Phase	
Species	$N =$	$\bar{X} \pm S.E.$	$N =$	$\bar{X} \pm S.E.$	"t"$=$[a]
Antigonon leptopus	17	0.83 ± 0.08	13	0.65 ± 0.07	1.64
Cuphea aequipetala	8	0.62 ± 0.06	6	0.48 ± 0.11	1.27
Hymenocallis littoralis	11	26.71 ± 2.72	7	28.62 ± 7.63	0.28
Lamourouxia rhinanthifolia	15	1.19 ± 0.10	8	1.23 ± 0.22	0.19
Lobelia cardinalis	11	0.95 ± 0.23	12	2.64 ± 0.73	2.14*
	12	2.22 ± 0.40	9	3.48 ± 1.12	1.22[b]
Penstemon gentianoides	15	0.77 ± 0.17	11	1.65 ± 0.52	1.87
Penstemon barbatus	14	0.79 ± 0.10	12	0.91 ± 0.15	0.71
	12	0.77 ± 0.10	9	1.19 ± 0.10	2.21*
Delphinium nelsonii	27	0.21 ± 0.03	22	0.43 ± 0.12	2.72**
Cuphea llavea	9	4.25 ± 0.85	10	2.19 ± 0.85	U

Note: U = 71.   p < 0.025   Mann-Whitney U-test used.
[a]Staminate and pistillate phases compared with a "t" test.
[b]Nectar measured at 1030 h when hummingbirds first visited the flowers.
*p < 0.05   **p < 0.01

clude both phases. 2) Flowers from all parts of the inflorescence were sampled because the position of a flower in the inflorescence may influence the amount of nectar in a flower (Andrejeff 1932). 3) We sampled equal numbers of flowers, usually one, from each plant because interplant differences may be quite large (Andrejeff 1932; Cruden and Hermann 1982; Fahn 1949), and may be genetically controlled (Pedersen 1953; Walker et al. 1974). 4) The initiation of nectar secretion in a population is not synchronous. The sources of variation outlined above explain, at least in part, the large variation in our samples.

For most species, flowers were collected and the nectar was gently squeezed from the base of the corolla and collected in a micropipet. If the flowers had a long, thick floral tube, it was slit as the nectar was removed. Regardless, contamination of the nectar with cell sap was avoided.

Nectar volumes were measured in calibrated micropipets and sugar concentrations with Bellingham and Stanley pocket refractometers. The amount of sugar per flower (tables 3.1–3.5) is the product of the volume, the concentration, and a correction factor, and is given in milligrams of sucrose. The correction factor is determined from the regression equation $y = 0.0046x + 0.9946$ where x is the concentration (Cruden and Hermann 1982). Most nectars contain a mixture of sucrose, glucose, and fructose with minimal amounts of other sugars (Baker and Baker 1982; Percival 1962). Since the refractive indices of glucose and fructose are approximately one half that of sucrose (Hainsworth and Wolf 1972a), it is energetically meaningful, as well as convenient, to give the amount of sugar in sucrose units.

Sugar concentrations and the amount of sugar per flower may be overestimated by 8 to 11 percent due to the refractive properties of amino acids in the nectar (Inouye et al. 1980). This should not affect our general conclusions. For comparisons made within species we assume that the amounts of amino acids are relatively constant from population to population (Baker 1978; Baker and Baker 1976) and that there is little intra-populational variation. Comparisons made between the nectars of species with different classes of pollinators should not be affected because the magnitude of the differences is far greater than the error attributable to the presence of amino acids.

Rates of nectar secretion (figs. 3.2–3.4, 3.6) were determined by regressing amounts of accumulated nectar or sugar against time. Slopes that did not deviate significantly from zero are interpreted as evidence for lack of secretion (see fig. 3.5). The Mann-Whitney U-test and "t"-test were used where appropriate. Throughout the text means plus or minus the standard error are given.

Populations in appendix I were assigned to flower classes on the basis of floral characteristics (Faegri and Pijl 1979), time of flower opening, anthesis, and careful study of the behavior of flower visitors.

The study was conducted primarily in August and September 1973 to 1975. Unless otherwise indicated, all times are Mountain Standard Time.

RESULTS

Amount   Flowers pollinated by bats, birds, or hawkmoths produce relatively large volumes of nectar that contain large amounts of sugar compared to flowers pollinated by bees, butterflies, and small moths (fig. 3.1). Although the quantity is small, the concentration of

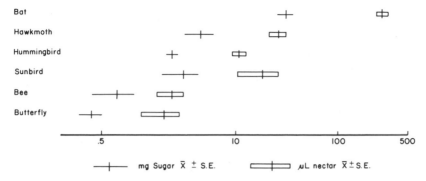

FIGURE 3.1
Amounts of sugar and nectar volumes per flower for flower classes in appendix I. $\overline{X}$ of $\overline{X}$'s ± S.E. are indicated on a log scale.

sugar is higher in the nectar of bee-pollinated flowers than in the nectar of either hawkmoth- or hummingbird-pollinated flowers (table 3.2).

Significant variation in sugar concentration within flower classes is associated with differences in elevation (table 3.2). The concentration of sugar in the nectars of hawkmoth-pollinated flowers from above 2400 m is lower than those from below 2400 m ($U = 83$; p $>$ 0.05). Likewise, both the amount of sugar and the volume of nectar per flower are lower in the high elevation flowers ($U = 108$; p $<$ 0.001 and $U = 100$; $p <$ 0.005, respectively). Sugar concentrations of hummingbird nectars also are significantly lower at high elevations ($U = 54$; $p = 0.05$). The amount of sugar per flower shows the same trend, but the difference is not statistically significant. We find no significant difference between the nectars of high and low elevation bee-pollinated flowers.

The sugar concentration of the nectar of some hummingbird-pollinated species increased during the morning (table 3.3). In *Penstemon kunthii* and *Hedeoma ciliolata* the increase in concentration was associated with a significant increase in volume and sugar content, an indication that active secretion occurred after 0630 h. In the other species the increase in concentration was not accompanied by a significant increase in either the volume or amount of sugar per flower. In *Salvia cardinalis*, *S. gregii*, and *Quamoclit coccinea*, species for

TABLE 3.2
COMPARISON OF NECTAR VOLUMES, AMOUNTS, AND PERCENT OF SUGAR IN FLOWERS ABOVE AND BELOW 2400 METERS

	$N=$	μl Nectar per Flower $\bar{X}$ of $\bar{X}$'s ± S.E.	mg Sugar per Flower $\bar{X}$ of $\bar{X}$'s ± S.E.	Percent Sugar $\bar{X}$ of $\bar{X}$'s ± S.E.
Hawkmoth Flowers				
above 2400 m	6	4.26 ± 0.05	0.80 ± 0.04	17.6 ± 0.7
below 2400 m	18	42.57 ± 12.95***	8.22 ± 2.29***	21.3 ± 1.1*
Hummingbird Flowers				
above 2400 m	8	9.00 ± 2.13	1.67 ± 0.46	18.3 ± 1.8
below 2400 m	9	9.77 ± 2.64	2.45 ± 0.62	23.8 ± 2.2*
Bee Flowers				
above 2400 m	5	1.52 ± 0.49	0.55 ± 0.21	29.0 ± 4.3
below 2400 m	7	2.51 ± 1.11	1.04 ± 0.52	34.9 ± 2.8

Note: $\bar{X}$'s compared with the Mann-Whitney U-test: *$p < 0.05$, ***$p < 0.001$

TABLE 3.3
DIURNAL CHANGES IN THE SUGAR CONCENTRATIONS OF NECTAR FROM HUMMINGBIRD-POLLINATED FLOWERS

Species	Time	N =	μl Nectar per Flower $\bar{X}$ ± S.E.	mg Sugar per Flower $\bar{X}$ ± S.E.	Percent Sugar $\bar{X}$ ± S.E.
Agastache cf pringlei	0630	9	2.37 ± 0.36	0.60 ± 0.09	24.7 ± 0.7
	1200	8	2.17 ± 0.35	0.62 ± 0.11	28.3 ± 1.5*
Castilleja integrifolia	0830	10	8.91 ± 1.92	1.80 ± 0.29	16.8 ± 0.8
	1320	6	13.56 ± 2.04	2.80 ± 0.46	20.4 ± 0.9**
Hedeoma ciliolata	0630	16	2.23 ± 0.34	0.35 ± 0.06	15.8 ± 0.2
	1500	16	3.75 ± 0.58*	0.64 ± 0.11*	17.2 ± 0.6*
Lamourouxia rhinanthifolia	0715	6	6.57 ± 0.88	1.35 ± 0.17	20.6 ± 0.2
	1115	11	4.72 ± 0.78	1.22 ± 0.17	27.1 ± 1.2***
Lobelia cardinalis	0645	23	18.79 ± 4.62	1.79 ± 0.43	9.3 ± 0.3
	1200	21	21.55 ± 3.46	2.61 ± 0.41	12.5 ± 0.6*
Penstemon kunthii	0630	19	2.82 ± 0.26	0.49 ± 0.04	17.7 ± 0.7
	1200	15	4.77 ± 0.60***	1.06 ± 0.13****	20.9 ± 1.0**
Salvia elegans	0735	6	8.15 ± 2.53	1.48 ± 0.40	19.1 ± 1.1
	1140	17	8.68 ± 1.61	2.15 ± 0.38	24.8 ± 0.4****

Note: $\bar{X}$'s compared with a "t" test: *p < 0.05, **p < 0.02, ***p < 0.01, ****p < 0.001.

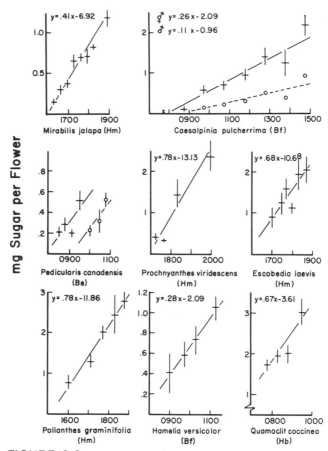

FIGURE 3.2

Rates of nectar accumulation in "fast" producers. $\overline{X} \pm$ S.E. are indicated by horizontal and vertical bars, respectively. Regressions are based on all data points (see appendix 1) and all are significant with a P <0.01. *Caesalpinia pulcherrima:* solid line = hermaphroditic flowers, open circles = male flowers, average of two measurements used for each point. Bf = butterfly; Hb = hummingbird; Hm = hawkmoth.

which we have comparable data, sugar concentrations did not change significantly from early to late morning. The relationship between sugar concentration and elevation in hummingbird-pollinated flowers was compared with the Spearman rank correlation coefficient and there is a significant negative correlation ($R_s = -0.504$; $p < 0.05$).

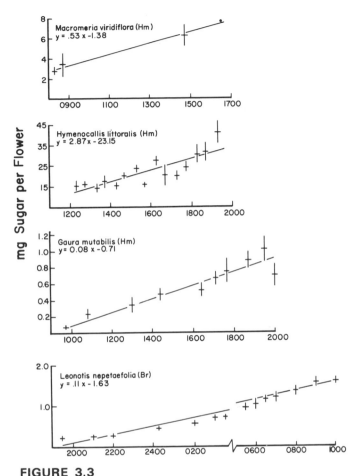

FIGURE 3.3
Rates of nectar accumulation in "slow" producers. $\overline{X} \pm$ S.E. are indicated. Regression equations of *Macromeria viridiflora* and *Hymenocallis littoralis* are based on all data points, those of other species on the $\overline{X}$'s. All regressions were significant with a P <0.01. Br = bird; Hm = hawkmoth.

Our data for a series of butterfly pollinated species suggest the amount of nectar in one large flower may be equivalent to that in many small flowers. The amount of sugar (0.42 mg) in the nectar of a *Lantana camara* inflorescence (4.2 $\pm$ 0.3 flowers) is equivalent to that in a flower of *Hamelia versicolor* or *Caesalpinia pulcherrima*, and can be reached from one place by a foraging butterfly. The same butter-

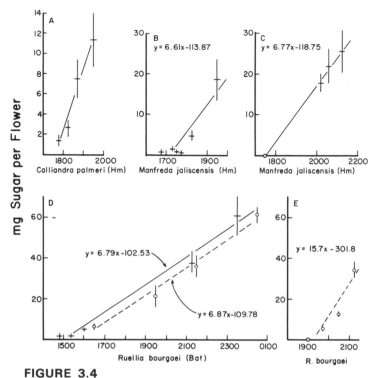

FIGURE 3.4

A-E. Rates of nectar accumulation in "super" producers. $\overline{X} \pm$ S.E. are indicated. Regression equations are based on all data points and all are significant with a P <0.01. C: measurements made on flowers from which nectar had been removed with a pipet and then bagged; D: solid line and crosses = 1976 data, slashed line and open circles = 1974 data; E: Measurements made on flowers from which nectar had been removed by hummingbirds and then bagged.

fly foraging on a *Tithonia* (Compositae) head has to change position, at least once, to reach all the open flowers ($\overline{X} = 41 \pm 4.4$ flowers per head).

Hawkmoth-pollinated plants from high elevations (>2400 m) compared to plants from lower elevations are farther apart and/or offer a smaller energetic reward (table 3.4). Flower densities of *Mirabilis jalapa*, a mid-elevation species, reach 65 per m², i.e., 83 mg sugar per m², compared to 0.50, 0.22, and 0.73 mg of sugar per m² in populations of *Gaura*, *Polianthes*, and *Prochnyanthes* (table 3.4), respectively.

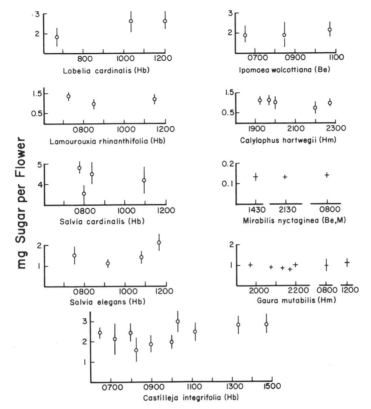

FIGURE 3.5

Species in which there was not significant nectar secretion following initiation of activity by coevolved pollinators. Hummingbirds (Hb) initiated activity ca 0630 h: hawkmoths (Hm) ca 1900 h; bees (Be) ca 0800 h. The regression equation for each species did not vary significantly from zero. *Lobelia cardinalis:* The lower value at 0630 is due to lower amounts of nectar in newly opened flowers. Hummingbirds initiated foraging at 1015 and no nectar accumulated thereafter.

($y = -0.01x + 2.77$, $n = 55$, $r^2 = 0.00$); *Ipomoea wolcottiana:*
($y = 0.08x + 1.65$, $n = 30$, $r^2 = 0.01$); *Lamourouxia rhinanthifolia:*
($y = 0.02x + 1.47$, $n = 25$, $r^2 = 0.01$); *Calylophus hartwegii:*
($y = -0.03x - 1.75$, $n = 54$, $r^2 = 0.01$); *Salvia cardinalis:*
($y = -0.05x + 5.19$, $n = 21$, $r^2 = 0.002$); *Mirabilis nyctaginea:*
($y = 0.001x + 0.126$, $n = 10$, r^2 0.04); *Salvia elegans:*
($y = 0.24x - 0.64$, $n = 44$, $r^2 = 0.07$); *Guara mutabilis:* open circles = 24 Aug. 1973 ($y = 0.01x + 0.79$, $n = 36$, $r^2 = 0.001$), closed circles = 3–4 Aug 1974 ($y = 0.02x + 0.49$, $n = 24$, $r^2 = 0.03$); *Castilleja integrifolia:* open circles = 5 Sep 1974 ($y = 0.15x + 0.99$, $n = 37$, $r^2 = 0.07$), closed circles = 6 Sep 1974 ($y = -0.15x + 3.64$, $n = 27$, $r^2 = 0.03$).

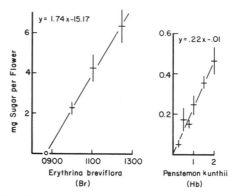

FIGURE 3.6

Resumption or continuation of nectar secretion following nectar removal. Nectar was removed from the flowers of *Erythrina breviflora* with pipets between 0830 h and 0845 h. Nectar was removed from *Penstemon kunthii* flowers by hummingbirds. Inflorescences were bagged, samples collected and measurements made throughout the day. Samples were collected 15, 30 and 120 minutes following removal. X's are based on sample sizes of approximately thirty.

Significant differences in nectar accumulation were found between flowers in the staminate and pistillate phases in protandrous *Delphinium nelsonii* and *Cuphea llavea* (table 3.1), as well as, between the male and hermaphroditic flowers of *Caesalpinia pulcherrima* (appendix I and fig. 3.2).

Rates and Initiation   The species we studied can be placed into one of three classes with respect to the rate of nectar secretion, i.e., slow, fast, and super. The slow producers (fig. 3.3) secrete 5 to 10 percent of their maximum accumulation per hour in contrast to fast producers (fig. 3.2) that secret 22 to 68 percent of their maximum per hour. In one hour super producers (fig. 3.4) secret as much nectar as a fast producer secrets in two to three hours. In *Manfreda brachystachya*, a super producer, a period of slow secretion may precede the period of rapid secretion. Secretion may begin as early as 1400 h but relatively little accumulation occurs prior to 1800 h.

The time of initiation of nectar secretion reflects the protection the nectar receives and the activity period of the pollinators. The nectar

TABLE 3.4
DISTANCE BETWEEN AND ENERGETIC REWARD PER
FLORAL UNIT

Species	N =	Number Flowers per Unit	mg Sugar per Unit[a]	Distance Between Units (meters)
Gaura mutabilis[b]	21	1.98 ± 0.22	1.86	3.29 ± 0.74
Polianthes durangense[b]	13	2.25 ± 0.22	1.43	6.27 ± 1.97
Prochnyanthes mexicana[b]	6	3.50 ± 0.45	3.33	4.55 ± 1.41
Manfreda brachystachya	35	3	6.36	1.17 ± 0.22

Note: Floral unit = the number of flowers that a pollinator can visit from one position.
[a]Number of flowers per floral unit × milligrams of sugar per flower (see appendix I).
[b]High elevation species.

of slow producers is protected by a thick corolla or calyx. The nectar of many fast producers is more liable to illegitimate removal. Initiation of secretion is so timed that sufficient amounts of nectar are present when the pollinators become active. For example, many hawkmoth-pollinated flowers initiate nectar secretion between 1600 and 1800 h, 2 to 4 hours prior to hawkmoth activity (fig. 3.2).

Cessation and Resumption In most species nectar is secreted at a constant rate (figs. 3.2–3.4, 3.6) until some critical amount has accumulated, and then ceases (fig. 3.5). Cessation occurs prior to or shortly after the initiation of pollinator activity. Evidence of cessation was obtained in four additional species. The sampling times and amounts of sugar are given: *Macrosiphonia macrosiphon:* 1740 h— 5.32 ± 1.00 mg; 1945 h—5.58 ± 0.96 ($t = 0.19$, $df = 11$). *Leonotis nepetaefolia:* 0850 h—0.96 ± 0.06 mg; 1045 h—0.92 ± 0.06 mg ($t = 0.500$, $df = 20$). *Datura ceratocaula:* 1640 h—14.72 ± 2.97 mg; 1935 h—11.92 ± 2.63 mg ($t = 0.674$, $df = 7$). In *Hamelia versicolor* nectar secretion probably ceased in early afternoon: 1640 h—1.65 ± 0.25 mg, n = 8; 1855 h—1.42 ± 0.15 mg, n = 14 ($t = 0.852$).

We found two classes of exceptions to the time of cessation. First, in a few species nectar secretion ceased well after the activity of the pollinators reached its daily maximum, e.g., in *Hamelia versicolor* and *Quamoclit coccinea*. Second, in *Caesalpinia pulcherrima* nectar secretion was continuous.

The third element in the typical pattern is the resumption or continuation of secretion if nectar is removed (figs. 3.4C, 3.4E, 3.6). In *Manfreda brachystachya* and *Penstemon kunthii* accumulation was constant and in the former was equivalent to that in bagged flowers. We cannot explain the large difference in the rates of nectar secretion in robbed and unvisited flowers of *Ruellia bourgaei* (figs. 3.4D and 3.4E). Resumption or continuation of secretion was observed in other species including *Penstemon barbatus*, *Calliandra callistemon* and *Macromeria viridiflora*.

In contrast to the above species, nectar secretion did not resume following removal by flower visitors in *Salvia reflexa* and *Quamoclit coccinea*. Flowers of *S. reflexa* (n = 40) were marked and bagged immediately after being visited by honey bees (*Apis mellifera*); no flowers contained nectar two hours later. In the case of *Q. coccinea*, nectar secretion began at approximately 0730 h and continued through 0930 h (fig. 3.2) but flowers bagged at 1300 h, then examined at 1400 and 1550 h, contained no measurable amounts of nectar.

Our data from *Penstemon kunthii* suggest that nectar secretion ceases at approximately the same time that hummingbirds cease daily activity, i.e., ca 1830 h, and that little or no secretion occurs during the period of hummingbird inactivity. First, flowers bagged following visits by hummingbirds between 1630 and 1720 h then measured one, two, and three hours later contained equivalent amounts of nectar, i.e., 0.30 ± 0.04 mg (n = 20); 0.30 ± 0.04 mg (n = 19); and 0.35 ± 0.06 mg (n = 6) of sugar, respectively. Second, on a different night the amount of sugar available per flower at 2130 h ($\overline{X} = 0.22 \pm 0.04$ mg; n = 22) was equivalent to that in flowers between 0300 and 0430 h the following morning ($\overline{X} = 0.30 \pm 0.05$ mg; n = 23) ($t = 1.143$). Likewise, the volumes of nectar were equivalent. Nectar secretion resumed in the morning between 0430 and 0500, one and one-half to two hours prior to the initiation of hummingbird activity. Nectar secretion in *P. gentianoides* has a similar pattern. No nectar accumulates during the night (table 3.5) and secretion starts in the morning, prior to the beginning of bee activity. Although we have no data as to when secretion starts and stops in other species, it is clear that many protandrous flowers produce nectar in both the staminate and pistillate phases (table 3.1). Our data suggest that *Cuphea llavea* is an exception. One

TABLE 3.5
NOCTURNAL CHANGES IN NECTAR IN FLOWERS OF
PENSTEMON GENTIANOIDES

Day and Time	N =	µl Nectar per Flower $\overline{X} \pm S.E.$	Percent Sugar $\overline{X} \pm S.E.$	mg Sugar per Flower $\overline{X} \pm S.E.$
24 Aug 2000	13	3.28 ± 0.70	46.7 ± 2.7	2.27 ± 0.76
25 Aug 0630	16	3.43 ± 0.54	29.6 ± 1.9	1.50 ± 0.37
25 Aug 1800	15	5.84 ± 1.06	32.1 ± 1.3	2.31 ± 0.54
26 Aug 0645	13	5.22 ± 0.83	26.4 ± 0.9	1.64 ± 0.28

half of the pistillate phase flowers examined contained no nectar and those that did contained significantly less nectar than flowers in the staminate phase (table 3.1). We suggest that nectar found in the pistillate phase flowers was produced in the staminate phase and overlooked by the hummingbirds.

Resorption Large plants of *P. gentianoides*, whose flowers remain open for three days, were bagged after cessation of bumblebee activity and nectar accumulation was monitored over the next two days. The volume of nectar per flower did not change during either night (table 3.5) and there was an equivalent decrease in the amount of sugar each night. This resulted in a dramatic (17 percent) decrease in sugar concentration during night one compared to night two (5.7 percent). Both decreases were significant ($t = 5.28$, $p < 0.001$; and $t = 3.35$, $p < 0.01$; respectively). The difference in the decrease in concentration between the two nights must be a consequence of the greater volumes of nectar in the flowers at the end of day 2, a result of bumblebees being excluded from the flowers during the day. Amounts of nectar in the staminate and pistillate phase flowers were equivalent (table 3.1). The decrease in sugar concentration in the two types of flowers was also equivalent.

Environmental Influence Local climatic conditions may affect nectar secretion in several ways. The initiation of nectar secretion in a population of *Pedicularis canadensis* following a chilly, dewy night was approximately one and one-half hours later compared to a morning after a warm, dewless night (fig. 3.2). The rate of nectar secretion on the two days was equivalent. In *Caesalpinia* the rate of secretion seems to be affected by temperature. The mean rate of secretion from 0830 to 1030 h (0.24 $\pm$ 0.01 mg sugar/hr; n = 5 days) is significantly lower than the rate of secretion from 1030 to 1230 h (0.37 $\pm$ 0.03 mg sugar/hr; n = 3 days), t = 4.85, p < 0.01). For the sample periods temperatures in the shade increased from 29° to 32° to 32° to 36°C, respectively. A sample of flowers collected at noon, which were still wet, and presumably cooler, contained 0.47 $\pm$ 0.07 mg of sugar per flower, about 30 percent of the sugar normally present in flowers at that time.

DISCUSSION

Our data corroborate the observations of other workers that flowers pollinated by bats, hawkmoths, and birds produce substantially more nectar and provide a greater energetic reward than flowers pollinated by bees, butterflies, and other low-energy requiring insects (table 3.6). Unfortunately, much of the data in Fahn (1949) and Percival (1965) were not reported in such a manner as to be comparable with ours and, of necessity, have been omitted from table 3.6.

TABLE 3.6
RANGES AND AVERAGE AMOUNTS OF SUGAR CONTAINED IN THE NECTAR OF FLOWERS IN SEVERAL FLOWER CLASSES

Flower Class	Number Species	mg Sugar per Flower Range of $\overline{X}$'s	mg Sugar per Flower $\overline{X}$ of $\overline{X}$'s ± S.E.
Bat	5[a]	10.2–43.6	26.5 ± 6.4
Hawkmoth	17	0.66–26.4	5.02 ± 1.67
Hummingbird	18[a,b]	0.31–12.0	2.90 ± 0.62
Sunbird	9[a,c]	0.19–9.19	3.44 ± 0.96
Bee	63[d,e]	0.0024–6.19	0.69 ± 0.16
Butterfly	10[a,f]	0.024–0.73	0.39 ± 0.08

[a]Includes some data from Percival (1965).
[b]Includes *Salvia* (Hainsworth and Wolf 1972a).
[c]Includes *Aloe graminicola* (Wolf 1975).
[d]Includes data from Heinrich (1975a, c, 1976).
[e]Includes data from Hocking (1968).
[f]Includes *Poinciana regia* (Fahn 1949).

The concentration of sugar in the nectar of many, if not most, plants also reflects the gustatory preferences or physical limitations of the pollinators. The nectars of bee-pollinated flowers tend to have higher sugar concentrations than the nectars of flowers pollinated by other animals (Baker 1975) and honeybees, at least, prefer sugar concentrations of 30 to 50 percent (Waller 1972). The nectars of most butterfly-pollinated flowers fall within the range of 15 to 25 percent (Watt, Hoch, and Mills 1974; appendix I) and models developed by Kingsolver and Daniel (1979) suggest that the rate of nectar extraction is mechanically limited and that concentrations of 20 to 25 percent optimize the net energy gain by the butterflies. In contrast, the sugar concentrations of the nectars of hummingbird-pollinated flowers are less than those preferred by hummingbirds (Hainsworth and Wolf 1976; Stiles 1976). Bolten and Feinsinger (1978) suggested that the lower than preferred concentrations are an adaptation that discourages foraging by bees. Earlier, Baker (1975) argued that the lower sugar concentrations reduced the viscosity of the nectar such that it could be easily and rapidly removed from the flowers by the hummingbirds. We return to the viscosity argument below.

Within each flower class, there is substantial variation in nectar volume and amount of sugar available to pollinators when they begin foraging. This variation reflects a number of factors, including discrete subgroups within a flower class, differences in latitude and elevation, number of flowers per floral unit, and distance between floral units.

Flower classes need not be uniform with respect to nectar characteristics; for example, we recognize three groups of bird-pollinated flowers: hummingbird-, sunbird-, and "oriole/starling"-pollinated flowers. The nectars of sunbird and hummingbird flowers have equivalent sugar concentrations and volumes but the former tend to be sucrose-poor whereas the latter tend to be sucrose-rich (Baker and Baker 1982, pers. comm.). Further, they differ significantly in the concentration and numbers of amino acids in their nectars (Baker and Baker 1973b, pers. comm.).

Foraging time budgets of the few hummingbirds and sunbirds that have been studied are equivalent (Gill and Wolf 1975; Wolf 1975; Wolf, Hainsworth, and Gill 1975), thus it is not surprising that the energetic rewards produced by hummingbird- and sunbird-pollinated flowers are equivalent. In contrast to hummingbird and sunbird flowers, "oriole/starling" flowers, e.g., *Spathodea campanulata* and *Erythrina breviflora*, produce large volumes of weak nectar that contain large amounts of amino acids and/or proteins (Baker and Baker, pers. comm.; Cruden and Toledo 1977; Toledo and Hernandez 1979). The nectars of these flowers are sucrose-poor, there being little, if any, sucrose in the species studied (Baker and Baker, pers. comm.; Cruden and Toledo 1977; Feinsinger et al. 1979; Skead and Niven 1967). The energetic demands of these large passeriform birds must be quite different from those of hummingbirds and sunbirds and this is reflected in the volumes and constituents of the nectars of the flowers they visit and pollinate. Likewise, flowers pollinated primarily by large bees, e.g., carpenter bees and bumblebees, should produce more nectar than flowers pollinated by small solitary bees or honey bees.

Differences in habitat may contribute to within-class variation in available nectar and its sugar content. Andrejeff (1932), Bonnier and Flahault (1878), and Hocking (1968) have reported that bee flowers from high elevations and latitudes produce nectars that offer a greater energetic reward than conspecifics at lower elevations and latitudes. Based on such observations, Heinrich and Raven (1972) argued that because pollinators expend more energy at low temperatures, it is reasonable to expect that flowers pollinated at low temperatures will provide greater energetic rewards than those that are pollinated at higher temperatures. We detected no such pattern in our work, and in the one case where a comparison of conspecifics can be made, i.e., *Cuphea aequipetala*, flowers from the lower elevation population provide a considerably larger energetic reward than flowers in the high elevation popualtion. The flowers of the lower elevation population were visited by hummingbirds, as well as by bees, and it has been suggested that such populations may be shifting to hummingbird pollination (Cruden 1976a).

In contrast to the nectars of bee-pollinated flowers, the nectars of hawkmoth- and hummingbird-pollinated flowers in high elevation habitats have lower sugar concentrations than those from lower elevations (Baker 1975; Hainsworth 1973; Hainsworth and Wolf 1972b; table 3.2). Both hawkmoths and hummingbirds hover while extracting nectar from a flower. The rate of extraction is a function of the nectar's viscosity, which increases exponentially with decreased temperature (Baker 1975). Thus, in a "cold" habitat, maximum energetic profit will be realized from less viscous nectars, i.e., nectars with lower sugar concentrations and faster handling times.

Our evidence suggests that sugar concentration as it affects the viscosity of nectar in hummingbird-pollinated flowers, at least over elevational gradients and diurnally in high elevation species, is more important than the discouragement of foraging bees. First, there is a significant negative correlation between elevation and concentration and no correlation with corolla length (the measure of accessibility to bees used by Bolten and Feinsinger, 1978) in the species we studied ($R_s = 0.360$) (appendix I; unpub. data). Second, in six of the seven high elevation species the sugar concentration increased between early and late morning (table 3.3) as it did in one of three low elevation species.

We suggest that the flowers actively regulate the sugar concentration of their nectar such that the viscosity remains relatively constant during the day. Small changes in either the amount of water and/or sugar in the nectar would significantly alter the sugar concentration and thus its viscosity. For species that are still secreting nectar an increase in the concentration can be achieved by altering the concentration of the nectar secreted later in the morning. For those flowers that have "ceased" secretion other mechanisms, e.g., active resorption of water and/or secretion of sugar must account for the changes in concentration. Evaporation is not a satisfactory explanation because the corollas of some species are quite thick and in only one species is there a noticeable decrease in the volume of the nectar. Further, in those species in which the sugar concentration of the nectar increased during the morning, the change is such as to make the nectar more attractive to foraging bees. The critical experiments that will resolve the question of "viscosity" vs. "discouragement" remain to be done and it is quite possible that low sugar concentrations reflect different selection pressures.

That it is adaptive for a flower to be able to alter the sugar concentration of its nectar is seen also in *Catalpa speciosa* Warder. The nectar has a sugar concentration of 23.7 percent (vol. = 3.23 μl) at sunrise and 39 percent (vol. = 2.56 μl) at sunset (Stephenson and Thomas 1979). The data suggest that the flowers produce a nectar adapted for moths at night, and by decreasing the quantity of water in the nectar during the day, secrete a nectar that is more attractive to bees. Both moths and bees are important pollinators of this species.

Our data for hawkmoth-pollinated flowers offer a sharp contrast with the predictions of Heinrich and Raven (1972) that plants in "cold" habitats will provide a greater energetic reward than equivalent plants in warmer habitats. Compared to several mid-elevation species, individuals of high elevation species are farther apart and/or provide a smaller energetic reward per floral unit (tables 3.2 and 3.4). The latter is a consequence of lower sugar concentrations and smaller volumes. Similar differences were found between high and low elevation populations of *Calliandra anomala* (Cruden 1976a). Because the hawkmoths are active for quite short periods of time (Cruden et al. 1976) only a limited number of flowers can be visited compared to flowers in more hospitable habitats, which are visited repeatedly. In cold habitats, maximum fecundity should result from maximizing fruit set rather than seed set because the initial pollinator visit to a flower results in greater seed set than any subsequent visit to that flower (Silander and Primack 1978). Thus, any adaptation that forces a pollinator to visit increased numbers of flowers should be selectively advantageous. Low nectar production is such an adaptation (Cruden 1976b). (See appendix II for discussion of why lower elevation hawkmoth-pollinated flowers produce greater amounts of nectar.)

A third source of within flower-class variation is derived from the spacing of flowers, i.e., the number of open flowers per plant and the distance between plants. Heinrich (1975a) presented data which show that in Compositae and other plants whose flowers occur in aggregates, small amounts of nectar in many flowers are equivalent to large amounts of nectar in a few flowers. A comparison of nectar available per floral unit in butterfly-pollinated species further illustrates the point. Likewise, the amounts of sugar available per inflorescence of bat-flowered *Agave palmeri* and *A. lecheguilla* are of the same magnitude, even though the flowers of the latter species produce almost four times as much sugar as those of *A. palmeri* (appendix I).

The distance between floral units should also be reflected in the amount of nectar produced per flower, i.e., the energetic reward should be proportional to the energy expended to reach the reward (Heinrich and Raven 1972) and our observations (Cruden and Hermann, unpubl.) suggest this is the case. The relationship of distance between floral units to nectar production merits further study.

A typical flower begins to secrete nectar prior to the activity of its pollinators. The rate of secretion is constant and continues until some critical amount has accumulated and then nectar secretion ceases. Nectar secretion resumes only if nectar is removed. We have documented all these phases and other workers have described one or more aspects of the pattern, including: a constant rate of secretion (Hainsworth and Wolf 1972a; Stiles 1975; Wolf 1975); cessation once some maximum is reached (Boëtius 1948; Hainsworth and Wolf 1972a; Stiles 1975); and resumption of secretion if nectar is removed (Boëtius 1948; Raw 1953; Wykes 1950).

The beginning of nectar secretion reflects the activity period of the pollinator class and, in some cases, the degree to which the nectar is protected from nectar thieves. In general, nectar secretion begins one to four hours before the coevolved pollinators become active (fig. 3.2). Butterfly-pollinated flowers of *C. pulcherrima* and *Hamelia versicolor* began nectar secretion approximately one and one-half hours before butterfly activity reached its maximum. In *Penstemon kunthii* nectar secretion resumed one and one-half to two hours prior to hummingbird activity as it did in two species studied by Feinsinger (1976). Also, most of the hawkmoth-pollinated species initiated nectar secretion one to three hours prior to the activity period of the hawkmoths. A small number of hawkmoth-pollinated flowers began nectar secretion ten or more hours before the hawkmoths became active and in each instance the nectar was protected by a thick corolla and/or calyx.

Although our data suggest that maximum nectar accumulation occurs prior to or shortly after initiation of pollinator activity, there are exceptions to this rule. The first includes flowers that initiate nectar secretion shortly before pollinator activity and if unvisited reach a maximum well after the initiation of pollinator activity, e.g., various species of *Heliconia* (Stiles 1975), *Penstemon kunthii*, and *Hamelia versicolor*. The second group includes species that initiate nectar secretion well after their pollinators have become active and present their nectar out of phase with other species that have the same pollinator class. The flowers of hummingbird-pollinated *Rigidella flammea* (Iridaceae) open in the late afternoon compared to congenerics (Cruden 1971) and other hummingbird-pollinated plants whose flowers open in early morning.

Differences in the time of nectar secretion and/or presentation can play a role in organizing activities of a pollinator guild. In protandrous, temporally dioecious umbels, e.g., *Sium suave* Walt., *Cicuta maculata* L., *Pastinaca sativa* L., and probably *Heracleum lanatum* Michx., plants in the staminate phase produce nectar one to two hours prior to plants in the pistilate phase (Cruden and Sharon Ward, unpub.). We suggest the difference in timing of nectar secretion maximizes the efficiency of relatively inefficient pollinators, i.e., flies and wasps. This difference in the timing of nectar presentation is sufficient to allow the pollinators to acquire large pollen loads prior to their first visits to pistillate phase plants.

Asynchronous timing of nectar production in habitats where there is competition for pollinators would be adaptive, because staggering of nectar production would tend to decrease competition for pollinators temporally. Such would appear to be the case in three Compositae that flower simultaneously in alpine meadows in Colorado. Based on pollinator behavior and the presence of nectar in their flowers, *Erigeron simplex*, *Senecio crocatus*, and *Erigeron peregrinus* have maximum amounts of nectar at different times during the day, i.e., mid-morning, late morning, and early afternoon, respectively (Cruden and I. Baker, unpub.). Kleber (1935) presented data that suggests various bee-pollinated flowers reached maximum accumulation at different times. Likewise, two species of butterfly-pollinated *Anguria* produce nectar at different times of the day but it is not clear if this is a mechanism to avoid competition for pollinators or maintain reproductive isolation (Gilbert 1975).

Cessation   Cessation of nectar production may occur at two periods in the daily cycle of secretion, first, once some maximum is reached, and second, in several day flowers, during the period of pollinator inactivity. It is clear that hummingbird-pollinated *Penstemon kunthii* and bumblebee-pollinated *P. gentianoides* do not secrete while their pollinators are inactive. Lack of secretion during the pollinator's inactive period also occurs in *Fritillaria* (Beutler 1930), *Citrus*, *Tecomaria* (Fahn 1949), *Inga*, *Cuphea* (Feinsinger 1976), *Penstemon barbatus* and *Ipomopsis aggregata* (Brown and Kodric-Brown 1979).

The reverse pattern, nocturnal secretion and diurnal cessation occurs in *Agave schottii* (Schaffer et al. 1979). The pattern, volumes and rates of nectar secretion reported suggest hawkmoth pollination. In *Asclepias verticillata* secretion occurs primarily between 1800 and 2200 h (Willson, Bertin, and Price 1979) and is correlated with the activity of noctuid moths. As in the case of *A. schottii* the flowers of *A. verticillata* are visited by diurnal visitors which effect significant amounts of pollen transfer.

Recently, Frankie, Opler, and Bawa (1976) reported discontinuous nectar secretion in *Andira inermis*, a self-incompatible tree. Nectar "flowed" from 0730 to 0830 h and again from 1100 to 1400 h. They attributed no adaptive significance to this unusual pattern.

We wish to emphasize the fact that nectar secretion may cease after some maximum is reached. If nectar secretion is not continuous then measuring the nectar of flowers bagged for twenty-four hours will give a gross underestimate of the amount of nectar that can be produced during a day. For example, *Erythrina breviflora* flowers can produce 1.74 mg of sugar per hour; over a twelve-hour day this is more than four times the amount of sugar in unvisited flowers.

Resorption   Active resorption of nectar may not be common. Even though nectar may be energetically expensive to make, it must also be expensive to actively resorb. The flowers of *P. gentianoides* resorb enough sugar from their nectar between dusk and dawn to reduce the concentration 17 percent. During this period their pollinators are inactive. The population studied was at 3850 m and at the time bumblebees became active temperatures were 7° to 9°C. We suggest, in this case, that resorption of the sugar reduces the viscosity of nectar such that early foraging bumblebees can take nectar quickly and efficiently. Rapid handling of the nectar would act to increase the number of flowers visited, hence increase fecundity. This is particularly important in habitats subject to significant amounts of rain and/or low temperatures each day, as are those in which *P. gentianoides* is frequently found. Temporal reduction in nectar concentration as in *Echium vulgare, Sinapsis alba* (Boëtius 1948), and *P. gentianoides*, is not to be confused with consistently low nectar concentrations typical of the nectars of high elevation hummingbird- and hawkmoth-pollinated flowers. Resorption of nectar in pollinated and/or old flowers has been reported in *Platanthera* (Bonnier 1878), *Echium vulgare, Rubus idaeus*, (Boëtius 1948), and *Trifolium repens* (Pankratova 1950). (See also appendix III.)

We view several studies reporting resorption of nectar with some reservation, i.e., Pedersen et al. (1958), Shuel (1961), Ziegler and Lüttge (1959). Their evidence of resorption is based on the substitution of C^{14}-sucrose or C^{14}-glutamic acid for the nectar and demonstrating that labeled sugars or amino acids appear in the flower and elsewhere in the plant. These studies showed that nectar constituents move into the surrounding tissue but failed to demonstrate that there was a net movement of constituents from the nectar into the plant. (See also appendix IV.)

Alternative Patterns A comparison of alternative patterns of nectar secretion suggests that the pattern of available nectar is adaptively significant. Feinsinger (1978) reported that hummingbirds visited more artificial flowers when they contained variable amounts of "nectar" compared to when they contained uniformly large amounts of "nectar." Substantial variance in the available nectar results from a number of factors operating singly or in combination, including within and between plant differences in the rates and amount of nectar produced per flower, the age of the flowers, the pattern of nectar secretion, the pattern of flower opening, and pollinator foraging.

In dichogamous flowers the amount of nectar produced may change with the age of the flower. In some protandrous species, for example in *Delphinium*, the older pistillate phase flowers produced substantially more nectar than staminate phase flowers. In protogynous *Metrosideros collina* (Forster) Gray nectar production was greatest at the time of stigma receptivity (Carpenter 1976), i.e., in younger flowers. In these species the differences in amounts of nectar in pistillate and male phase flowers are tied to the breeding systems of the plants.

Considerable variance in available nectar may result from other patterns of secretion. Feinsinger (1978) discussed five species in which approximately half of the flowers contained little or no nectar, a small number contained a great deal of nectar, and the remainder contained variable amounts of nectar. Our data from *Cuphea llavea* show that such a pattern may result from protandrous flowers producing nectar only in the staminate phase with pistillate phase flowers containing only what remained from the staminate phase. Some of Feinsinger's (1978) data suggest that the same pattern may also result from flowers on the same plant producing different amounts of nectar.

Continuous opening of flowers during the period of pollinator activity will produce an equivalent pattern. In *Monarda fistulosa* L. the flowers open continuously, 7 to 9 percent per hour, from 0800 to 2000 h (Cruden, Hermanutz, and Shuttleworth, unpub.). The quantity of nectar in just opened flowers is much greater than in older, visited flowers.

Significant variance in the amount of nectar per flower may be typical of most species and Feinsinger's (1978) experiment provides a selective explanation. In some plants, for example, *Delphinium* and *Cuphea llavea*, differences in nectar secretion are clearly genetic in nature as is the pattern of flower opening in *Monarda*. The large differences in nectar per flower between plants of *Ipomopsis aggregata* (Cruden and Hermann 1982; fig. 3.2) also may have a genetic basis. Other differences are clearly environmental in nature but contribute to the pattern. For example, pollinators do not visit each flower on the plants they visit, and those that are missed or visited early in the foraging period will contain large amounts of nectar relative to those visited later in the foraging period.

We have encountered two additional patterns. First, in *Caesalpinia pulcherrima* nectar secretion is continuous and is the key to its reproductive biology (Cruden 1976a; Cruden and Hermann-Parker 1979). Very briefly, where pollinator activity is high, each flower receives many visits, and where pollinator activity is low, nectar accumulates in the flowers and the butterflies spend longer periods of time at each flower. The longer visits compensate for the smaller number of visits because pollen transfer is proportional to foraging time. This is a consequence of pollination being effected through wing-stigma contact and the butterflies fluttering continuously while foraging.

Second, in *Fagopyrum esculentum* (Beutler 1930), *Medicago sativa* (Pedersen and Bohart 1953), *Salvia reflexa*, and *Quamoclit coccinea* nectar secretion does not resume following removal by a flower visitor. In each instance, one visit by a pollinator is probably sufficient to effect some cross-pollination and/or sufficient selfing to maximize seed set. Although the breeding systems of the four species are dissimilar, they share the attribute that additional visits by a pollinator will probably not increase seed set, and continued nectar secretion would be energetically wasteful. Based on its pollen-ovule ratio (i.e., 1245:1) *Salvia reflexa* is facultatively xenogamous, yet its fecundity (98 percent) is equivalent to that of species that regularly self-pollinate. *Q. coccinea* is self-compatible (Doctors van Leeuwan 1938) but requires a pollinator. Because the stigma is adjacent to the anthers, hummingbirds, when they probe the flowers for nectar, must push the anthers into the stigma, thus effecting self-pollination. Many races of *F. esculentum* are self-compatible and self-pollination may be effected, in spite of the species being distylous. With respect to *M. sativa*, once the flowers are tripped and presumably pollinated, nectar secretion ceases (Pedersen and Bohart 1953). Visits to tripped flowers do not result in additional pollen transfer.

Rates We have divided flowers into three groups based on the relative amount of nectar they secrete per hour. Associated with rate of secretion are the time of initiation and protection of the nectar. Slow producers are early initiators and have thick corollas or calyces which provide protection for the nectar. Conversely, fast producers initiate secretion later and their flowers offer less protection to the nectar. The super producers include species whose flowers are regularly "robbed" by hummingbirds one-half to one and one-half hours prior to initiation of pollinator activity. Yet, when their pollinators become active the flowers of super producers contain amounts of nectar equivalent to those in species whose flowers are not robbed. We suggest that the high rates of nectar secretion are adaptations that permit these species to be competitive for pollinators, in spite of the regular loss of nectar to robbers. Two additional species, *Macromeria viridiflora* and *Prochnyanthes mexicana*, and a population of *Calliandra anomala* (Cruden 1976a, b), which occur in high elevation habitats, are regularly "robbed" by hummingbirds. Although the rates of secretion in flowers of these populations are low relative to those of the super producers of low elevations, the amounts of nectar available to the hawkmoths when they become active are equivalent to those of other hawkmoth-pollinated species of high elevations. Both *M. viridiflora* and *P. mexicana* begin secretion well in advance of hawkmoth activity and continue that rate of secretion after hummingbird visits. They are not super producers.

Nectar Production As a Component of Breeding Systems
We have already discussed the role of nectar production in the reproductive biology of *Salvia reflexa* and *Quamoclit coccinea*, i.e., they produce enough nectar to attract a pollinator, once.

In dichogamous species and species with unisexual flowers, different amounts of nectar may be produced at different stages of development or may vary with the sex of the flower. The flowers of *Delphinium nelsonii* are strongly protandrous and mature acropetally, thus flowers in the staminate phase occur higher on the scape than those in the pistillate phase. The latter flowers produce nearly twice as much nectar as the staminate flowers (0.43 vs. 0.21 mg of sugar respectively). One of us (SP) watched more than one hundred foraging trips of *Bombus* to inflorescences of *D. nelsonii* and in only three instances did the bees work down the scape (see also Epling and Lewis 1952). *Bombus* forage in a similar manner on protandrous *Digitalis purpurea* (Bierzychudek and Best, pers. comm.; Manning 1956) and *Chamaenerion* (= *Epilobium*) *angustifolium* (Benham 1969). In the former species, the lower pistillate phase flowers produced more nectar than the upper staminate phase flowers, and Bierzychudek (1977) suggested, as do we, that the behavioral patterns of the bees maximize cross-pollination in such self-compatible species, while the bees maximize their foraging efficiency.

In other species differences in nectar production may influence the relative amounts of time spent by pollinators at a flower. In protogynous *Metrosideros collina* nectar production is highest at the time of stigma receptivity (Carpenter 1976). The larger volume of nectar should result in longer visits and increased pollination. In *Caesalpinia pulcherrima*, hermaphroditic flowers produce approximately twice as much nectar as male flowers. The nectar in the male flowers is found in the nectary at the base of a tubular petal. This adaptation brings the butterfly into contact with the anthers, but little more. In contrast, nectar rises up the petal tube of the hermaphroditic flowers. The greater volume of nectar keeps the butterflies at the flowers for a longer period of time, thus increasing the likelihood of pollination (Cruden 1976a; Cruden and Hermann-Parker 1979). Similar differences in nectar volumes and sugar content have been reported in various monoecious and dioecious species (Bawa and Opler 1975; Fahn 1949; Perkins et al. 1975). The sex with the largest amount of nectar varies with the species. For example, pistillate flowers of *Musa velutina* produce less nectar than the staminate flowers (Percival 1965). We suggest that the interaction between the pollinators and the flowers producing the larger amounts of nectar should be proportional to the time spent at the flower, whereas the interaction with the flowers producing less nectar should be independent of time spent at the flower. The interaction may result in the transfer of pollen to a pollinator or to a stigma.

Environmental Effects Our studies suggest that both temperature and rain affect nectar secretion. Low temperatures may delay the start of nectar secretion or decrease the rate of secretion. In *Caesalpinia pulcherrima*, wet flowers in the shade contained less nectar than those that had been in direct sunlight for a period of time. Further, the rate of nectar secretion between 0830 and 1030 h was lower than that from 1030 to 1230 h. The gradual increase in nectar secretion (Cruden 1976b) undoubtedly reflects the gradually increasing temperature during the periods of observation. Diurnal changes in temperature might well explain increased rates of nectar secretion in *Ipomopsis aggregata* (Brown and Kodric-Brown 1979) and *Erythrina fusca* (Feinsinger et al. 1979). In field studies, it is difficult at best, to segregate effects due to light and those due to temperature.

The literature on the effect of rain and/or high humidity on nectar secretion and its quality is confusing and much of it is difficult to place in any perspective (see Bonnier 1878; Buetler 1930, 1953; Hocking 1968; Kleber 1935; Shuel 1952). Most workers agree that the concentration of nectar increases and its volume decreases as a consequence of evaporation. Conditions favoring evaporation frequently occur at mid-day or early afternoon when relative humidities are low and temperatures high. Likewise all workers agree that rain water can wash away or dilute nectar.

High humidity also affects nectar quality. Buetler (1930), Boëtius (1948), and Kleber (1935) reported that the volume of nectar was greater at high humidities but that the concentration of sugar was less. They attribute this to absorption of water from the atmosphere. In contrast, in *Calliandra anomala* the amount of sugar in the nectar on a rainy day was decreased but the volume of the nectar was equivalent to that in the flowers on a dry day (Cruden 1976a).

Differences in nectar production may reflect differences in soil moisture. We found striking differences in the volumes and concentrations of nectar in male flowers of *Caesalpinia pulcherrima* near Miramar, Colima, in different years: 1974—1.62± 0.26 μl, 23.7± 2.6 percent, and 0.42± 0.08 mg sugar; 1976—0.77± 0.14 μl, 47.5± 4.4 percent, and 0.43± 0.05 mg sugar per flower, respectively; n = 6 for both samples. The same plants were sampled and the temperatures were equivalent and the humidities high. The condition of the vegetation suggested that the region had received less rain in 1976. A similar pattern was reported in *Ipomopsis aggregata* by Waser (1978). Mean 24 h production in 1975 and 1976 was 1.04 mg sugar ad 6.5 μl of nectar and 1.08 mg sugar and 3.42 μl of nectar per flower, respectively.

It is evident that the coevolution between plants and their pollinators are manifest in a number of ways. The selective impact of the pollinators on the plants is seen in the amounts of nectar produced and to a lesser extent the timing of nectar production. On the other hand, the plants, through regulation of nectar production, can effectively channel the behavior of the pollinators while maximixing their fecundity with a minimal energetic expenditure.

APPENDIX I

VOLUME, NECTAR CONCENTRATION, AND SUGAR CONTENT OF NECTAR (PER FLOWER) AT TIME OF FIRST VISIT BY COEVOLVED POLLINATORS

Species and Locality[a]	N =	μl Nectar $\overline{X} \pm S.E.$	Percent Sugar $\overline{X} \pm S.E.$	mg Sugar $\overline{X} \pm S.E.$
BAT-POLLINATED				
Agave palmeri Engelm.[b]				
Arizona: Cochise Co. 3 mi W Portal	5	64.4 ± 8.3	17.6 ± 1.5	11.0 ± 2.24
Agave cf. *lecheguilla*[c]				
Durango: Rt 40, ca 20 km NE Pedricena	5	149.3 ± 21.9	12.9 ± 2.3	43.1 ± 7.2
Ipomoea muricoides Roem. & Schult.				
Oaxaca: nr Oaxaca, rd to Monte Alban, 1880 m	13	161.2 ± 16.8	27.3 ± 0.7	48.9 ± 4.9
Ruellia bourgaei Hemsl.				
Jalisco: Rt 45, K 18-19, N Guadalajara, 1300 m (2154)	38(4)[d]	182.5[e]	19.9	40.2[e]
HAWKMOTH-POLLINATED				
Calliandra anomala (Kunth.) Macbr.				
Morelos: Rt 95D, 1.5 km N rd to Cuautla, 2300 m (2084)	46(6)[d]	39.2[e]	14.1	6.3[e]
Calliandra palmeri S. Wats.				
Jalisco: Rt 15, K 94, 16 km NW Magdalena, 1210 m (2066)	17(4)[d]	69.3[e]	16.2	10.68[e]
Calylophus hartwegii (Benth.) Rowen				
Chihuahua: Rt 45, ca 21 km E Parral, 1600 m (2028)	10	5.96 ± 0.94	24.3 ± 0.6	1.53 ± 0.17
Durango: Rt 30, 58-59 km E Rt 45, 2110 m (2301)	54	4.32 ± 0.34	24.2 ± 0.4	1.13 ± 0.09
Castilleja sessiliflora Pursh				
Iowa: Dickinson Co., Freda Haffner Preserve	7	6.93 ± 0.68	17.2 ± 0.9	1.27 ± 0.12
Castilleja mexicana (Hemsl.) Gray				
Durango: Rt 30, 58-59 km E Rt 45, 2120 m (2133)	14	3.83 ± 0.63	24.8 ± 0.9	1.04 ± 0.17
Datura meteloides A. DC.				
Arizona: Cochise Co. just E of Portal	3	49.85 ± 5.88	29.2 ± 2.7	15.69 ± 0.81

Locality	n			
Datura ceratocaula Ort.				
Chihuahua: 16 km NW Alvaro Obregon, 1900 m	7	68.44 ± 7.98	18.7 ± 1.2	13.52 ± 1.95
Datura stramonium L.				
Iowa: Johnson Co., Rt 218 4 mi W Iowa City	4	83.58 ± 5.37	21.1 ± 0.6	17.78 ± 1.27
Escobedia laevis Cham. & Schlecht.				
Jalisco: Rt 80, 1–2 km S rd to Ayutla, 1330 m (2071)	19	10.85 ± 0.92	17.5 ± 0.3	2.02 ± 0.16
Gaura mutabilis Cav.				
Durango: Rt 40, 54–55 km W Durango, 2550 m (2045A)	20	4.39 ± 0.27	19.5 ± 0.7	0.94 ± 0.09
D. F.: Rt 95D, K 26, ca 25 km S Tlalpan, 2480 m (2236)	8	4.45 ± 0.71	15.4 ± 1.0	0.77 ± 0.15
Hymenocallis littoralis Salisb.				
Sinaloa: Rt 40, K 268, ca 22 km E Rt 15, 20 m (2054)	56(16)[d]	213.3	20.1	37.7
Macromeria viridiflora DC.				
Durango: Rt 40, 45–46 km W Durango, 2400 m (2183)	8	4.19 ± 1.25	19.2 ± 1.3	0.88 ± 0.27
Macrosiphonia macrosiphon (Torr.) A. Heller				
Durango: Rt 30, ca 18 km W Rt 45, 1820 m (2035B)	17	13.67 ± 1.40	31.9 ± 0.5	4.89 ± 0.51
Manfreda brachystachya (Cav.) Rose				
Jalisco: Rt 15, K 94, 16 km NW Magdalena, 1210 m (2065)	5	124.5 ± 27.9[e]	16.3 ± 0.5	19.79 ± 0.5[e]
Mirabilis jalapa L.				
Jalisco: Rt 15, below Anones, ca 41 km NW Magdalena, 1060 m	79(16)[d]	5.35	24.8	1.27
Oenothera hookeri Torr. & Gray				
New Mexico: Alto Co. 4 mi W Alto (2004B)	11	4.87 ± 0.96	24.1 ± 1.9	1.26 ± 0.28
Oenothera laciniata Hill				
Durango: Rt 40, 52–54 km W Durango, 2550 m (2185)	14	4.13 ± 0.96	17.0 ± 1.6	0.76 ± 0.20
Oenothera stubbei Dietrich & Raven				
Nuevo Leon: rd to Doctor Arroyo, 20 km S of rd to Linares, 2150 m (2221)	12	41.24 ± 5.88	17.7 ± 0.6	7.39 ± 1.26
Polianthes durangensis Rose				
Durango: Rt 40, 53–54 km W Durango, 2550 m	20	4.18 ± 0.60	15.7 ± 0.6	0.66 ± 0.12
Polianthes graminifolia Rose				
Jalisco: Rt 54, ca 14 km N Guadalajara, 1400 m (2069)	8	13.45 ± 1.11	21.0 ± 0.8	3.03 ± 0.23

APPENDIX I (*cont.*)

Species and Locality[a]	N =	μl Nectar $\bar{X}$ ± S.E.	Percent Sugar $\bar{X}$ ± S.E.	mg Sugar $\bar{X}$ ± S.E.
Prochnyanthes mexicana (Zucc.) Rose				
Durango: Rt 40, 53–54 km W Durango, 2550 m (1854)	8	12.02 ± 2.07[e]	18.1 ± 1.1	2.28 ± 0.38[e]
	26(18)[f]	4.24 ± 1.22	18.5 ± 0.6	0.81 ± 0.23
Ruellia cf. *metzae* Tharp				
Coahuila: Rt 57, K 216–217, ca 10 km NE Nava, 250 m (2125)	11	7.54 ± 0.74	20.6 ± 0.4	1.68 ± 0.17
HUMMINGBIRD-POLLINATED				
Agastache cf. *pringlei* (Briq.) Luit.				
Durango: Rt 40, 55–56 km W Durango, 2530 m (2044)	9	2.37 ± 0.36	24.7 ± 0.7	0.60 ± 0.10
Castilleja tenuifolia Mart. & Gal.				
Jalisco: Rt 110, ca 5 km N Mazamitla, 2060 m	32	6.42 ± 0.43	18.6 ± 0.5	1.29 ± 0.10
Castilleja integrifolia L.f.				
Durango: Rt 40, K 131, ca 30 km W El Salto, 2750 m (2000)	10	8.91 ± 1.92	16.8 ± 0.8	1.80 ± 0.29
Cuphea llavea Lav. & Lex.				
Jalisco: Rt 110, ca 5 km N Mazamitla, 2060 m (2240)				
pistillate phase	10	7.24 ± 2.30	25.5 ± 1.5	2.19 ± 0.85
staminate phase	9	14.53 ± 2.46	25.8 ± 1.0	4.25 ± 0.85
Hedeoma ciliolata (Epling and Stewart) Irving				
Nuevo Leon: 20 km S of rd to Linares on rd to Doctor Arroyo, 2150 m (2222)	16	2.23 ± 0.34	15.8 ± 0.2	0.35 ± 0.06
Lamourouxia rhinanthifolia HBK				
Durango: Rt 40, 54–55 km W Durango, 2550 m (2045)	6	6.57 ± 0.88	20.6 ± 0.2	1.35 ± 0.17
Rt 40, 46 km W Durango, 2420 m	23	7.46 ± 0.62	15.9 ± 0.4	1.26 ± 0.10
Lobelia cardinalis L.				
Mexico: 5 km NE Sta. Marta, rd to Tres Cumbres, 3110 m (2203)	23	18.79 ± 4.62	9.3 ± 0.3	1.79 ± 0.43

Operculina alatipes (Hook.) House				
Sinaloa: Rt 40, 1.5 km E Rt 15, 10 m (2195)	9	9.33 ± 1.70	35.7 ± 0.4	4.01 ± 0.77
Penstemon kunthii G. Don				
Durango: Rt 40, 45–46 km W Durango, 2400 m	20	4.95 ± 0.36	17.2 ± 0.7	0.91 ± 0.07
Penstemon barbatus Nutt.				
Durango: Rt 40, 45–46 km W Durango, 2400 m	21	5.24 ± 0.38	17.4 ± 0.4	0.94 ± 0.06
Psitticanthus caylculatus (DC.) G. Don				
Jalisco: Rt 110, ca 5 km N Mazamitla, 2060 m (2239)	6	25.32 ± 4.12	18.2 ± 1.5	5.26 ± 1.39
Quamoclit coccinea (L.) Moench				
Jalisco: Rt 80, ca 10 km N rd to Ayutla, 970 m (2070)	20(4)[d]	5.07	27.7	1.18
Salvia cardinalis Kunth				
Michoacan: Mirador Atzimba, ca 53 km E Morelia, 2940 m (2081)	21	17.71 ± 1.28	24.6 ± 0.7	4.71 ± 0.38
Salvia elegans Vahl.				
Durango: Rt 40, K 131, ca 30 km W El Salto, 2750 m (2145)	6	8.15 ± 2.53	19.1 ± 1.1	1.48 ± 0.40
Salvia gregii Gray				
Nuevo Leon: rd to Doctor Arroyo, 20 km S rd to Linares, 2150 m (2216)	17	3.49 ± 0.29	29.7 ± 1.1	1.15 ± 0.09
Salvia pubescens Benth.				
Mexico: 6–7 km NE Tingambato, 1450 m (2170)	6	20.72 ± 4.43	24.2 ± 1.3	5.10 ± 1.06
SUNBIRD-POLLINATED				
Leonotis nepetaefolia R. Br.				
Sinaloa: Rt 40, 35 km E Rt 15, 30 m	34	5.94 ± 0.17	15.2 ± 0.3	0.19 ± 0.03
Jalisco: Rt 45, ca 22 km N Guadalajara, 1220 m	9	13.72 ± 1.52	16.7 ± 0.4	2.50 ± 0.32
Rt 90, ca 48 km E Guadalajara, 1860 m	15	9.17 ± 0.83	16.2 ± 0.5	1.66 ± 0.17
Leonotis leonurus R. Br.				
California: Botanical Garden, Berkeley	4	10.92 ± 0.99	21.6 ± 2.3	2.59 ± 0.38
Phygelius aequalis Harv.				
California: Botanical Garden, Berkeley	4	57.35 ± 5.92	15.9 ± 0.6	10.88 ± 1.21
Tecomaria capensis Spach				
California: Botanical Garden, Berkeley	4	11.33 ± 0.77	18.4 ± 0.7	2.23 ± 0.12

APPENDIX I (*cont.*)

Species and Locality[a]	N =	μl Nectar $\overline{X}$ ± S.E.	Percent Sugar $\overline{X}$ ± S.E.	mg Sugar $\overline{X}$ ± S.E.
"ORIOLE-STARLING"-POLLINATED				
Erythrina breviflora DC				
Morelos: Rt 95 (Libre) ca 9 km N Cuernavaca, 1940 m (2175)	15	48.13 ± 3.92	10.4 ± 0.4	4.55 ± 0.35
Spathodea campanulata Beauv.				
Morleos: Trailer Park E Cuernavaca, 1520 m	8	651.7 ± 107.9	5.4 ± 0.9	41.3 ± 7.88
BEE-POLLINATED				
Convolvulus sepium L.				
Iowa: Iowa City nr High School	7	2.93 ± 0.25	23.6 ± 0.8	0.77 ± 0.07
Cuphea aequipetala Cav.				
Mexico: 5 km NE Sta. Marta, rd to Tres Cumbres, 3100 m (2231)	14	1.42 ± 0.16	34.1 ± 0.9	0.56 ± 0.06
Jalisco: Rt 110, ca 5 km N Mazamitla, 2060 m (2241)	14	7.39 ± 1.31	43.3 ± 3.4	3.63 ± 0.59
Delphinium nelsonii Greene				
Colorado: pistillate phase	22			0.43 ± 0.12
staminate phase	27			0.21 ± 0.03
Ipomoea wolcottiana Rose				
Oaxaca: nr Oaxaca, below Monte Alban, 1880 m (2261)	13	5.64 ± 1.41	31.7 ± 1.1	2.21 ± 0.59
Minkelersia galactoides Mart. & Gal.				
Durango: Rt 40, K 132–133, ca 31 km W El Salto, 2760 m (2227)	23	2.35 ± 0.24	43.3 ± 1.0	1.24 ± 0.08
Oenothera deserticola (Loes.) Munz				
Durango: Rt 40, K 131, ca 30 km W El Salto, 2750 m (1244)	17	2.90 ± 0.11	22.7 ± 0.9	0.71 ± 0.14
Pedicularis canadensis L.				
Iowa: Cedar Co., Rochester Cemetery	32	0.95	35.5	0.42
Prunella vulgaris L.				
Durango: Rt 40, K 132–133, ca 31 km W El Salto, 2760 m	10	0.47 ± 0.07	20.4 ± 2.5	0.11 ± 0.02

Salvia tiliaefolia Vahl				
Nuevo Leon: rd to Doctor Arroyo, 20 km S rd to Linares, 2150 m	24	0.015	45.5	0.008
Salvia coahuilensis Fern.				
Nuevo Leon: rd to Doctor Arroyo, 20 km S rd to Linares, 2150 m (2214)	20	0.53 ± 0.06	32.6 ± 0.8	0.20 ± 0.02
Salvia reflexa Hornem.				
Nuevo Leon: rd to Doctor Arroyo, 20 km S rd to Linares, 2150 m (2215)	8	0.14 ± 0.03	32.4 ± 1.1	0.05 ± 0.01
Stachys radicans Epl.				
Mexico: 5 km NE Sta. Marta, rd to Tres Cumbres, 3100 m. (2232)	8	0.44 ± 0.15	24.6 ± 3.1	0.11 ± 0.03
BUTTERFLY-POLLINATED				
Antigonon leptopus H. & A.				
Sinaloa: Rt. 40, 1.5 km E Rt 15, 10 m	30	4.02 ± 0.26	17.2 ± 0.3	0.75 ± 0.06
Caesalpinia pulcherrima Sw.				
Sinaloa: Rt 40, 28 km E Rt 15, 30 m				
hermaphroditic flowers	47(7)[d]	1.22	20.0	0.28
male flowers	9(5)[d]	0.77	5.0	0.04
Hamelia versicolor Gray				
Colima: Rt 200, ca 16 km W Santiago, 20 m (2163)	24(4)[d]	2.52	16.9	0.42
Lantana camara L.				
Nayarit: 0.5 km W Rt 15 on rd to Est. Tetitlan, 640 m	16	0.51 ± 0.07	18.3 ± 0.7	0.10 ± 0.02
Nyctaginea capitata Choisy				
Durango: Rt 45, K 201 ca 39 km S Ent. La Zarca, 2060 m (2140)	24	3.00 ± 0.23	24.8 ± 0.4	0.81 ± 0.06
Tithonia cf. *macrophylla* Wats.				
Colima: above Rio Tuxpam, ca 7 km SE Colima, 530 m (2245)	9	0.078 ± 0.019	30.0 ± 1.1	0.027 ± 0.007
BEE- AND BUTTERFLY-POLLINATED				
Grovonia scandens L.				
Colima: above Rio Tuxpam, ca 7 km SE Colima, 530 m (2246)	16	0.58 ± 0.05	49.5 ± 4.3	0.33 ± 0.04

APPENDIX I (cont.)

Species and Locality[a]	N =	μl Nectar $\bar{X}$ ± S.E.	Percent Sugar $\bar{X}$ ± S.E.	mg Sugar $\bar{X}$ ± S.E.
SMALL MOTH-POLLINATED				
Gaura coccinea L.				
Durango: Rt 30, 58–59 km E Rt 45, 2110 m (2136)	17	1.71 ± 0.22	28.9 ± 3.6	0.31 ± 0.04
SMALL BEE AND SMALL MOTH-POLLINATED				
Mirabilis nyctaginea (Michx.) MacM.				
Iowa: Dickinson Co., Iowa Lakeside Laboratory	18	0.57 ± 0.02	23.5 ± 0.9	0.13 ± 0.01

[a] The number in parenthesis at the end of most localities is the collection number of RWC. Vouchers will be deposited at UC.

[b] 197 flowers open on branch, ca 2.1 gms of sugar available per branch.

[c] 45 flowers open per scape, can 1.9 gms of sugar per plant.

[d] Estimates of nectar volumes and amount of sugar per flower derived from regression analysis: Total number of flowers, followed by number of sample periods in parenthesis.

[e] Maximum available if flowers are not robbed by hummingbirds (see text for discussion).

[f] Amount of nectar available at time of hawkmoth activity; nectar concentrations based on 18 samples.

APPENDIX II

During the discussion, the question was raised as to why hawkmoth-pollinated flowers from lower elevations produced more nectar than flowers from high elevations, i.e., why couldn't they get by on far less nectar? We offer two suggestions. The first is easily tested and the second is speculative.

First, at lower elevations the moths exist continuously in a warm to hot environment. The metabolic rates of insects at mid (1800 m) and low (100 m) elevations must be significantly higher than conspecifics at high elevations (<2400 m) and we suggest that they utilize greater amounts of food reserves while inactive. Hawkmoths living at the lower elevations are active for longer periods of time and thus expend more energy in this facet of their daily cycle. Further, hawkmoths at high elevations are inactive during the coldest part of the night and it follows that their energy expenditure during that 4–6 hour period will be minimal while moths at lower elevations are expending a maximum amount of energy. That moths in warmer habitats expend greater amounts of energy is supported, at least in part, by Hanegan and Heaths' 1970 observations (*J. Exp. Biol.* 53:611–627) that moths are active for longer periods of time and expend more energy at high temperatures. Their experimental evidence with respect to activity parallels our observations in the field (Cruden et al. 1977). From these observations we suggest that hawkmoths living at higher elevations actually expend less energy than those living at lower elevations and that the energetic reward produced by the flowers reflects the differences in the energy budget of moths living at different elevations.

We hasten to point out, that the adaptive value of producing small energetic rewards in high elevations is not invalidated if the above argument is true.

The second argument assumes that at some point in time all hawk-moth-pollinated plants produced equivalent amounts of nectar and that in the mid or low elevations a "cheater," i.e., a species producing more nectar, gained a competitive advantage by keeping moths at the flowers longer, thus increasing its fecundity relative to that of competing species. Selection would then favor greater nectar production in the other species. Eventually the selective advantage of greater nectar secretion would be balanced by decreasing fecundity that would result from visitor satiation, i.e., the pollinators could obtain sufficient nutrient from one or a few flowers to satisfy their energetic requirements and would leave many flowers unvisited.

We offer a set of observations, from our work with *Manfreda brachystachya*, that shows a correlation between high rates of nectar secretion and high pollinator activity on the one hand and low rates of secretion and low activity on the other. The observations were made during the evenings of August 7, 8, and 10, 1973, ca 92 Km WNW of Guadalajara, Jalisco. On the first two nights the amount of sugar in bagged flowers at approximately 1930 h, shortly after the initiation of hawkmoth activity, was 18.5 ± 5.2 and 16.2 ± 2.0 mg of sugar. These figures are roughly equivalent to rates of 4.1 and 4.6 mg sugar per hour. On the last night the comparable figures were 3.8 ± 0.5 mg of sugar and a rate of 0.96 mg of sugar per hour. On the first two nights we observed many moths and they were still active when we left the site at 2230 h. On the last night it was our impression that fewer moths were active at 1915 and an hour later very few moths were visiting the flowers. Environmental conditions were approximately the same on all evenings with the exception of August 8, during which there was a light rain that did not seem to affect hawkmoth activity. Regardless, the point can be made that of those insects that initiated foraging on August 10, few of them were still present an hour later and this change in behavior was correlated with a low rate of nectar production. Tentatively, we suggest that high rates of nectar secretion are necessary in *M. brachystachya* to elicit and maintain foraging activity by large numbers of hawkmoths.

APPENDIX III

The question was asked, "Might the sugar have been removed by microbial action?" Several reasons were provided as to why this was unlikely. First, Dr. Elias pointed out that nectar contains anti-microbial agents. Second, during the time period in question the temperatures ranged from 2 to 6°C. Most bacteria grow quite slowly at such temperatures. Third, for bacteria to have used the amount of sugar removed, the nectar should have been turbid. Even after sixty hours the nectar was quite clear.

APPENDIX IV

Herbert Baker stimulated us to reappraise the notion of resorption. We perceive three possible ways to explain the data of Ziegler and Lüttge (1961), etc. First, the movement of labeled sugars or amino acids from the nectar into the plant could be the result of random molecular movement with no net movement of the molecule class. This would be neither evolutionarily nor ecologically interesting. Second, the movement could be passive; the result of the movement of nutrients from various floral parts to the developing fruit thus creating a diffusion gradient that might include the nectar. Although such a physical response might result in the saving of energy rich compounds, it might not be the result of natural selection. Third, materials in the nectar might be actively resorbed. The utilization of energy to remove constituents from nectar is quite likely to be an adaptive response, as the amount of energy invested in nectar might be quite small, relative to the energy budget of the plant.

ACKNOWLEDGMENTS

We wish to thank Sharon Kinsman, Diana Loeb-Cruden, Blake Parker, Eillen Stanislav, Kathleen Sayre, Nancy Sayre, and Richard Sayre for aiding with the field work; and Rogers McVaugh, Peter Raven, Jerzy Rzedowski, Mario Sousa, John Strother, L. O. Williams, and Susan Verhoek for identifying various species. We are indebted to Bernd Heinrich for permitting us to use the raw data from his papers and to Herbert G. and Irene Baker for unpublished data on various bird-pollinated species. Herbert Baker's comments on an earlier draft were valuable in the preparation of this manuscript, as were the comments of Sue Blaisdell, Kathryn Grove, and Sharon Miller-Ward. The Bee Research Association provided translations of the papers by Boëtius and Pankratova and Diana Loeb-Cruden translated sections of the paper by Bonnier.

REFERENCES

Andrejeff, W. 1932. Ueber Nektarien und über die Menge des Nektars einiger Gehölzarten. *Mitt. Deutsch. Dendrol. Gesel.* 44:99–105.

Baker, H. G. 1975. Sugar concentrations in nectars from hummingbird flowers. *Biotropica* 7:37–41.

Baker, H. G. 1978. Chemical aspects of the pollination biology of woody plants in the tropics. In P. B. Tomlinson and M. H. Zimmerman, eds., *Tropical Trees as Living Systems*, pp. 57–82. Cambridge: Cambridge University Press.

Baker, H. G. and I. Baker. 1973a. Amino acids in nectar and their evolutionary significance. *Nature* (London) 241:543–545.

Baker, H. G. and I. Baker. 1973b. Some anthecological aspects of the evolution of nectar-producing flowers, particularly amino acid production in nectar. In V. H. Heywood, ed., *Taxonomy and Ecology*, pp. 243–264. New York–London: Academic Press.

Baker, H. G. and I. Baker. 1975. Studies of nectar-constitution and pollinator-plant coevolution. In L. E. Gilbert and P. H. Raven, eds., *Coevolution of Animals and Plants*, pp. 100–140. Austin: University of Texas Press.

Baker, I. and H. G. Baker. 1976. Analyses of amino acids in flower nectars of hybrids and their parents with phylogenetic implications. *New Phytol.* 76: 87–98.

Bawa, K. S. and P. A. Opler. 1975. Dioecism in tropical forest trees. *Evolution* 29:167–179.

Benham, B. R. 1969. Insect visitors to *Chamaenerion angustifolium* and their behavior in relation to pollination. *Entomologist* 102:221–228.

Beutler, R. 1930. Biologische-chemische Untersuchungen am Nektar von Immenblumen. *Zeit. Vergleich. Physiol.* 12:72–176.

Beutler, R. 1953. Nectar. *Bee World* 34:106–116, 128–136, 156–162.

Bierzychudek, P. 1977. Spatial patterns of nectar availability in *Digitalis purpurea* and the behavior of foraging bees. *Bull. Ecol. Soc. Amer.* 58(2):8.

Boëtius, J. 1948. Über den Verlauf der Nektarabsonderung einiger Blütenpflanzen. *Beih. Schweiz. Bienenztg.* 2:258–317.

Bolten, A. B. and P. Feinsinger. 1978. Why do hummingbird flowers secrete dilute nectars? *Biotropica* 10:307–309.

Bonnier, G. 1878. Les Nectaries. *Ann. Sci. Nat.*, Series 6, 8:5–212.

Bonnier, G. and C. Flahault. 1878. Observations sur les modifications des vegetaux suivant les conditions physiques du milieu. *Ann. Sci. Nat.*, Series 6, 7:93–125.

Brown, J. and A. Kodric-Brown. 1979. Convergence, competition, and mimicry in a temperate community of hummingbird-pollinated flowers. *Ecology* 60:1022–1035.

Carpenter, F. L. 1976. Plant-pollinator interactions in Hawaii: Pollination energetics of *Metrosideros collina* (Myrtaceae). *Ecology* 57:1125–1144.

Cruden, R. W. 1971. The systematics of *Rigidella* (Iridaceae). *Brittonia* 23: 217–225.

Cruden, R. W. 1976a. Intraspecific variation in pollen-ovule ratios and nectar secretion: Preliminary evidence of ecotypic adaptation. *Ann. Missouri Bot. Gard.* 63:277–289.

Cruden, R. W. 1976b. Fecundity as a function of nectar production and pollen-ovule ratios. In J. Burley and B. T. Styles, eds., *Tropical Trees: Variation, Breeding, and Conservation*, pp. 171–178. London and New York: Academic Press.

Cruden, R. W. and S. M. Hermann-Parker. 1979. Butterfly pollination of *Caesalpinia pulcherrima*, with observations on a psychophilous syndrome. *J. Ecology* 67:155–168.

Cruden, R. W. and V. M. Toledo. 1977. Oriole pollination of *Erythrina breviflora* (Leguminosae): Evidence for a polytypic view of ornithophily. *Plant Syst. Evol.* 126:393–403.

Cruden, R. W., S. Kinsman, R. Stockhouse II, and Y. Linhart. 1976. Pollination, fecundity, and the distribution of moth-flowered plants. *Biotropica* 8:204–210.

Doctors van Leeuwan, W. M. 1938. Observations about the biology of tropical flowers. *Am. Jard. Bot. Buitenzorg* 48:27–68.

Epling, C. and H. Lewis. 1952. Increase of the adaptive range of the genus *Delphinium*. *Evolution* 6:253–267.

Faegri, K. and L. van der Pijl. 1979. *The Principles of Pollination Ecology*. Oxford: Pergamon Press.

Fahn, A. 1949. Studies in the ecology of nectar secretion. *Palest. J. Bot.* (Jerusalem) 4:207–224.

Feinsinger, P. 1976. Organization of a tropical guild of nectarivorous birds. *Ecol. Monogr.* 46:257–291.

Feinsinger, P. 1978. Ecological interactions between plants and hummingbirds in a successional tropical community. *Ecol. Monogr.* 48:269–287.

Feinsinger, P., Y. B. Linhart, L. A. Swarm, and J. A. Wolfe. 1979. Aspects of the pollination biology of three *Erythrina* species on Trinidad and Tobago. *Ann. Missouri Bot. Gard.* 66:451–471.

Frankie, G. W., P. A. Opler, and K. S. Bawa. 1976. Foraging behaviour of solitary bees: Implications for outcrossing of a neotropical forest tree species. *J. Ecology* 64:1049–1057.

Gilbert, L. E. 1975. Ecological consequences of a coevolved mutualism between butterflies and plants. In L. E. Gilbert and P. H. Raven, eds., *Coevo-

lution of Animals and Plants, pp. 210–240. Austin: University of Texas Press.

Gill, F. B. and L. L. Wolf. 1975. Economics of feeding territoriality in the Golden-Winged Sunbird. *Ecology* 56:333–345.

Hainsworth, F. R. 1973. On the tongue of a hummingbird: Its role in the rate and energetics of feeding. *Comp. Biochem. Physiol.* 46A:65–78.

Hainsworth, F. R. and L. L. Wolf. 1972a. Energetics of nectar extraction in a small, high altitude, tropical hummingbird, *Selasphorus flammula. J. Comp. Physiol.* 80:377–387.

Hainsworth, F. R. and L. L. Wolf. 1972b. Crop volume, nectar concentration, and hummingbird energetics. *Comp. Biochem. Physiol.* 42A:359–366.

Hainsworth, F. R. and L. L. Wolf. 1976. Nectar characteristics and food selection by hummingbirds. *Oecologia* (Berlin) 25:101–113.

Heinrich, B. 1975a. The role of energetics in bumblebee-flower interrelationships. In L. E. Gilbert, P. H. Raven, eds., *Coevolution of Animals and Plants*, pp. 141–158. Austin: University of Texas Press.

Heinrich, B. 1975b. Energetics of pollination. *Ann. Rev. Ecol. Syst.* 6:139–170.

Heinrich, B. 1975c. Bee flowers: A hypothesis on flower variety and blooming times. *Evolution* 29:325–334.

Heinrich, B. 1976a. Resource partitioning among eusocial insects: Bumblebees. *Ecology* 57:874–889.

Heinrich, B. 1976b. The foraging specializations of individual bumblebees. *Ecol. Monogr.* 46:105–128.

Heinrich, B. and P. A. Raven. 1972. Energetics and pollination ecology. *Science* 176:597–602.

Hocking, B. 1968. Insect-flower associations in the high Arctic with special reference to nectar. *Oikos* 19:359–388.

Inouye, D. W., N. A. Favre, J. A. Lanum, D. M. Levine, J. B. Meyers, M. S. Roberts, F. C. Tsao, and Y.-Y. Wang. 1980. The effects of nonsugar nectar constituents on estimates of nectar energy content. *Ecology* 61:992–996.

Kleber, E. 1935. Hat das Zeitgedachtnis der Bienen biologische Bedeutung? *Zeit. Vergleich. Physiol.* 22:221–262.

Kingsolver, J. G. and T. L. Daniel. 1979. On the mechanics and energetics of nectar feeding in butterflies. *J. Theor. Biol.* 76:167–179.

Manning, A. 1956. Some aspects of the foraging behavior of bumblebees. *Behavior* 9:164–201.

Pankratova, N. M. 1950. Investigation of the process of secretion of nectar. (In Russian). *Zh. Obshch. Biol.* 11:292–305.

Park, O. W. 1929. The influence of humidity upon sugar concentration in the nectar of various plants. *J. Econ. Entomol.* 22:534–544.

Pedersen, M. W. 1953. Seed production in alfalfa as related to nectar production and honeybee visitation. *Bot. Gaz.* 115:129–138.

Pedersen, M. W. and G. E. Bohart. 1953. Factors responsible for the attractiveness of various clones of alfalfa to pollen-collecting bumblebees. *Agron. J.* 45:548–551.

Pedersen, M. W., C. W. LeFevre, and H. H . Wiebe. 1958. Absorption of C^{14}-labeled sucrose by alfalfa nectaries. *Science* 127:758–759.

Percival, M. S. 1962. Types of nectar in angiosperms. *New Phytol.* 60:235–281.

Percival, M. S. 1965. *Floral Biology*. London: Pergamon Press.

Perkins, G. P., J. R. Estes, and R. W. Thorp. 1975. Pollination of *Cnidoscolus texanus* (Euphorbiaceae) in south-central Oklahoma. *Southwest. Nat.* 20:391–396.

Raw, G. R. 1953. The effect on nectar secretion of removing nectar from flowers. *Bee World* 34:23–25.

Schaffer, W. M. and M. V. Schaffer. 1977. The reproductive biology of Agavaceae: I. Pollen and nectar production in four Arizona Agaves. *Southwest. Nat.* 22:157–168.

Schaffer, W. M., D. B. Jensen, D. E. Hobbs, J. Gurevitch, J. T. Todd, and M. V. Schaffer. 1979. Competition, foraging energetics, and the cost of sociality in three species of bees. *Ecology.* 60:976–987.

Scullen, H. A. 1940. Relative humidity and nectar concentration in fireweed. *J. Econ. Entomol.* 33:870–871.

Shuel, R. W. 1952. Some factors affecting nectar secretion in red clover. *Plant Physiol.* 27:95–110.

Shuel, R. W. 1961. Influence of reproductive organs on secretion of sugars in flowers of *Streptosolen jamesonii*, Miers. *Plant Physiol.* 36:265–271.

Silander, J. A. and R. B. Primack. 1978. Pollination intensity and seed set in the evening primrose (*Oenothera fruticosa*). *Amer. Midl. Nat.* 100:213–216.

Skead, C. J. and C. M. Niven. 1967. *The Sunbirds of Southern Africa: Also the Sugarbirds, the White-Eyes, and the Spotted Creeper*. Cape Town: Balkema.

Stephenson, A. G. and W. W. Thomas. 1979. Diurnal and nocturnal pollination of *Catalpa speciosa* (Bignoniaceae). *Syst. Bot.* 2:191–198.

Stiles, F. G. 1975. Ecology, flowering phenology, and hummingbird pollination of some Costa Rican *Heliconia* species. *Ecology* 56:285–301.

Stiles, F. G. 1976. Taste preferences, color preferences, and flower choice in hummingbirds. *Condor* 78:10–26.

Toledo, V. M. and H. M. Hernandez. 1979. *Erythrina oliviae:* A new case of oriole pollination in Mexico. *Ann. Missouri Bot. Gard.* 66:503–511.

Walker, A. K., D. K. Barnes, and B. Furgala. 1974. Genetic and environmental effects on quantity and quality of alfalfa nectar. *Crop Sci.* 14:235–238.

Waller, G. D. 1972. Evaluating responses of honey bees to sugar solutions using an artificial flower feeder. *Ann. Entomol. Soc. Amer.* 65:857–862.

Waser, N. M. 1978. Competition for hummingbird pollination and sequential flowering in two Colorado wildflowers. *Ecology* 59:934–944.

Watt, W. B., P. C. Hoch, and S. G. Mills. 1974. Nectar resource use by *Colias* butterflies. *Oecologia* (Berlin) 14:353–374.

Willson, M. F., R. I. Bertin, and P. W. Price. 1979. Nectar production and flower visitors of *Asclepias verticillata*. *Amer. Midl. Nat.* 102:23–35.

Wykes, G. R. 1950. Nectar secretion researches. *Aust. Beekeeper* 52:67–68.

Wolf, L. L. 1975. Energy intake and expenditures in a nectar-feeding sunbird. *Ecology* 56:92–104.

Wolf, L. L., F. R. Hainsworth, and F. B. Gill. 1975. Foraging efficiencies and time budgets in nectar feeding birds. *Ecology* 56:117–128.

Ziegler, H. and U. Lüttge. 1959. Über die Resorption con C^{14}-Glutaminsaure durch sezernierende Nektarien. *Naturwiss.* 5:176–177.

4

A BRIEF HISTORICAL REVIEW OF THE CHEMISTRY OF FLORAL NECTAR

HERBERT G. BAKER AND IRENE BAKER
UNIVERSITY OF CALIFORNIA, BERKELEY

Flowering plants that are pollinated by animals usually produce nectar in their flowers. Such nectar may be ingested directly by the flower visitor (as by birds, bats, Lepidoptera, and Diptera) or it may be carried back to a nest and used in the nourishment of larval insects (as by many Hymenoptera). In the latter case, it is usually concentrated and chemically altered into honey. Honey has long been collected by human beings from honeybee nests in the Old World; thus, there is a Mesolithic cave painting in Spain which appears to portray a woman gathering honey from a bees' nest (Coon 1955:421).

Although it has long been known that bees collect nectar from flowers and store honey in the nest, (Lorch (1978) points out that Aristotle was aware of this in the fourth century B.C.), it is only recently that the distinction has been made between honey, the derived product (with any sucrose largely converted into hexoses), and nectar that is present in the flower as a secretion from nectaries. During the nineteenth century, when so much progress was made in understanding the pollinatory relations of flowers, most authors refer to the liquid found in flowers as "honey" (e.g., as recent a publication as Müller 1883). Knuth (1906–9) has the modern usage of "nectar" for the floral product. The early history of the association of nectar with pollination is beautifully reviewed by Lorch (1978).

Nectar usually tastes sweet, so that its strong sugar content was recognized early (Caspary 1848, for example). The sugar-water constitution was well recognized by Müller (1883: 259–260) in the following percipient observation:

The loosestrife (*Lythrum salicaria* L.) is visited by a number of long-tongued flies, especially *Rhingia rostrata*. This fly, standing on one or more of the petals, after gently rubbing its fore feet together, and brushing its tongue and head with them, stretches its proboscis out to a length of 11–12mm. and thrusts it down into the flower, letting it remain there for from six to ten seconds. Immediately after withdrawing it from the tube, it manipulates one of the anthers with its labellae for a short time (one to two seconds) in order to add to the liquid non-nitrogenous food some solid nitrogenous matter in the shape of pollen-grains.

The French botanist Gaston Bonnier (1878) quoted analyses of the individual sugars present in some nectars, although his belief that they were there to assist the subsequent maturation of the seeds was a holdover view from the early nineteenth century that was generally discredited by that time (Baker and Baker 1982a). However, Kartashova (1968) suggests that excess nectar, not collected by visitors, may be resorbed by the gynoecium and materially affect seed maturation. Kartashova also suggests that nectar has an anti-infection role with antibiotic effects on possible disease agents. Other analyses followed, but it was not until the middle of the twentieth century that it was realized that many other organic substances besides sugar may be present in floral nectar. Beginning with the recognition of the presence of "organic acids" by Beutler (1930) and Niethammer (also 1930), a wide range of chemicals has been reported for various nectars. A list of the chemicals reported in the literature up to 1975 is presented by Baker and Baker (1975) (see also Lüttge 1977). Sugars, amino acids, proteins, lipids, antioxidants, alkaloids, phenolics, "vitamins," "organic acids," saponins, dextrins, and inorganic substances have been reported from nectars of various species, and the occurrence of amino acids is virtually, if not actually, universal (Baker and Baker 1973, 1975, 1977; Baker 1977, 1978).

The biological significance of the occurrence of these chemicals in nectar (both phylogenetic and anthecological) is only beginning to be worked out but, particularly for sugars and amino acids, the complements and concentrations can be seen to relate to the pollinator type for the flower species under consideration (Baker and Baker 1973, 1975, 1979, 1982a, b; Baker 1977, 1978) as well as to the taxonomic relations of the plants (Baker and Baker 1975, 1976, 1982a, b, c; Harborne 1977). Constancy of the amino acid and sugar complements of nectar, despite environmentally induced variations in volume and concentration of nectar, is demonstrated for amino acids by Baker and Baker (1977) and Kartashova (1965), and for sugars by Percival (1961) and Baker and Baker (1982a), although Walker et al. (1974) found some tendency to genetic and environmental changes in nectar of *Medicago sativa*. Consequently, it may be useful to draw attention to significant papers that have dealt with the various chemical components of nectar.

SUGARS

Sugar Complements of Nectars By far the largest number of chemical analyses have dealt with the sugars present in floral nectar. Studies on a large scale by Percival (1961) and Baker and Baker (1982a, b), and smaller-scale studies by Mauritzio (1959, 1962) and others, have shown that most nectars contain the disaccharide sucrose and the monosaccharides glucose and fructose in detectable quantities. The disaccharides maltose and melibiose, and the trisaccharides melezitose and raffinose are encountered less frequently. Some other sugars are rare or of dubious occurrence (Percival 1965; Baskin and Bliss 1969; Jeffrey, Arditti, and Koopowitz 1970; Gottsberger et al. 1973; Watt et al. 1974; Crane 1977, 1978; Baker and Baker 1982a). The proportions of the three frequently occurring sugars are rather constant within species but can show wide interspecific differentiation (Percival 1961; Baker and Baker 1979, 1982a, b). A review of the literature on specific differences in sugar complements is provided by Baker and Baker (1982a) (forty-nine cited major references, to which should be added listings by Kartashova 1965).

In our studies of nectar sugar contents we have distinguished those with sucrose/hexose ratios (by weight) of more than 0.999 as "sucrose-dominant," those with 0.5 to 0.999 as sucrose-rich," those with 0.1 to 0.499 as "hexose-rich," and those with less than 0.1 as "hexose-dominant."

Analyses of the sugars in nectars from 765 species by Baker and

Baker (1982a) have confirmed the observation by Percival (1961) that for certain families there is conservatism in the proportions of sucrose, glucose and fructose. Thus, the nectars of Lamiaceae and Ranunculaceae are characteristically sucrose-rich or sucrose-dominant, while nectars from the Brassicaceae and Asteraceae are dominated by hexose sugars. Other families (e.g., Fabaceae and Scrophulariaceae) show marked differences in the sugar ratios, even between closely related species (Baker and Baker 1982a, b, c). Harborne (1977) has also reported various sugar combinations that have taxonomic significance within the genus *Rhododendron* (Ericaceae). The "phylogenetic constraint" in the conservative families notwithstanding, there are detectable significant relations between the sugar content of floral nectar and the type of pollinating flower visitor that is being rewarded (Baker 1978; Baker and Baker 1979, 1982a, b, c). Table 4.1 provides a summary of our findings.

Nectar Sugar Ratio Correlations Percival (1961) commented upon the appearance of correlation between tubular flowers and sucrose richness and, by implication, a tendency for open-bowl flowers and hexose richness. Corbet (1979a) pointed out that hexose-rich nectars evaporate less readily, which would fit with their presence in bowl-shaped flowers.

Hummingbird flowers have high sucrose/hexose ratios (i.e., 0.5 and above), while those flowers primarily visited by passerine (perching) birds have low ratios (less than 0.499) (Baker and Baker 1979, 1982a, b, c). It is suggested that the "taste" for hexose sugars may have been developed from the use by many of the passerine birds of fruit juices that are generally hexose-dominated (Baker and Baker 1982a, b, and unpub.). By contrast, hummingbird flowers are often derived from large-bee flowers (Grant and Grant 1968: 84) which we have shown to be characteristically sucrose-rich or sucrose-dominated. Hummingbirds do not usually visit soft fruits. An observation of hummingbird utilization of a sugar source other than nectar was made by Southwick and Southwick (1980), but this was by ruby-throated hummingbirds on the phloem sap of birch trees (access to which had been made by the beaks of yellow-bellied sapsuckers). They suggest that this sap is rich in sucrose. However, hummingbirds

TABLE 4.1
NUMBERS OF SPECIES IN EACH OF FOUR SUGAR-RATIO
CATEGORIES ARRANGED BY PRINCIPAL POLLINATORS

	$\dfrac{S}{G+F}$						
	< 0.1	0.1 to 0.499	0.5 to 0.999	> 0.999	N	G [a]	P [b]
Overall	195	231	149	190	765	–	–
Hummingbirds	0	18	45	77	140	119.52	<.001
New World passerines	11	1	0	0	12	25.16	<.001
Sunbirds, etc.	24	9	2	0	35	28.07	<.001
Honeyeaters	18	4	0	0	22	36.87	<.001
Honeycreepers	5	1	0	0	6	10.57	<.02
Lorikeets, etc.	1	2	0	0	3	3.69	.30
Hawkmoths	2	8	19	32	61	41.16	<.001
Settling moths	3	14	11	15	43	70.07	<.001
Butterflies and skippers	5	17	24	29	75	24.23	<.001
Short-tongued bees + butterflies	23	21	3	0	47	38.07	<.001
Short-tongued bees	115	103	28	17	263	75.47	<.001
Long-tongued bees	13	75	49	66	203	42.40	<.001
New World bats	9	18	0	0	27	32.51	<.001
Old World bats	1	3	2	1	7	1.36	.90
Non-volant mammals	0	2	2	1	5	13.44	<.01
Wasps	2	7	4	5	18	1.24	.75
Beetles	1	3	2	3	9	1.22	.75
Flies	29	27	7	9	72	14.82	<.001

Source: Baker and Baker 1982a, b.
[a] G = G statistic (see Sokal and Rohlf 1969, ch. 16)
[b] P = probability of difference from *Overall*

will visit hexose-rich Old World bird flowers of suitable size, shape, and color when these are in cultivation, but the experimental studies of Stiles (1976) and Hainsworth and Wolf (1976) show that they have a preference for nectar rich in sucrose.

Flowers visited nocturnally tend to have hexose-rich nectar if the visitors are New World bats of the suborder Microchiroptera, tend to be richer in sucrose if the visitors are Old World bats of the suborder Megachiroptera, and are usually strongly sucrose-rich or dominant if the visitors are hawkmoths (Sphingidae) (Baker and Baker 1982a, b). Similar sucrose-richness to that of the hawkmoth nectars characterizes flowers visited nocturnally by settling (i.e., not hovering) moths and by diurnally flying butterflies and skippers (Baker and Baker 1979, 1982a, b; see also I. Baker in Cruden and Hermann-Parker 1979). However, in the alpine zone of the Colorado Rockies, there is a less than optimum environment for sugar production by photosynthesis, and a majority of the plants have energy-economical hexose-rich nectar (Baker and Baker, unpub.). Some of them are visited by *Colias* butterflies which Watt et al. (1974) have shown have the enzymic capability of interconverting hexose and sucrose, so that they can make use of nectars with any combination or proportion of the sugars.

Bee flowers may be divided into those adapted to "short-tongued" bees (with tongues less than 6mm in length) and those adapted to "long-tongued" bees. The former group of visitors (together with flies) are rewarded with nectars exhibiting a wide range of sugar ratios, with a prevalence of hexose-dominance. On the other hand, the "long-tongued" bee flowers produce nectars that are usually sucrose-rich or dominant (Baker and Baker 1982a, b). Consideration of these results of analyses must include the fact that certain families characterized by hexose-richness of their nectar (e.g., Brassicaceae and Asteraceae) contribute heavily to the short-tongued bee data while other families (e.g., Lamiaceae and Ranunculaceae) characterized by sucrose-rich nectar contribute many species to the long-tongued bee flower group. It is not easy to distinguish cause from effect in these relationships.

A particularly interesting historical case concerns the nectar of "honey-bee-flowers". In 1952, Wykes (1952a, b) concluded from experiments that honeybees prefer a "balanced" nectar, with roughly equal quantities of sucrose, glucose and fructose. Percival's survey (1961) showed such *balanced* nectars to be rather uncommon (and that was also found in our survey) and Waller (1972) and Bachman and Waller (1977) performed feeding experiments that showed honeybees to prefer sucrose-rich solutions of sugars. These bees, which stand almost on the threshold between short-tongued and long-tongued bees, may *prefer* sucrose-richness but, in nature, will certainly take nectar that is overbalanced on the side of the hexoses.

Flowers that are visited by both short-tongued bees and butterflies (including many Asteraceae and Brassicaceae with flat-topped inflorescences that provide a "standing platform" and opportunities for walking between flowers) have hexose-richness or dominance (Baker and Baker 1982a, b). Wasp flower nectars seem to be richer in sucrose than those of short-tongued bee flowers (Baker and Baker 1982a, b), but for beetles there are not yet enough data for conclusions to be drawn.

These generalizations for different flower-pollinator groups have been examined on a finer taxonomic scale for species of *Penstemon* (Scrophulariaceae) (contrasts between hummingbird- and hymenopteran-pollinated species), *Campsis* (Bignoniaceae) (hummingbird and sunbird flowers), *Erythrina* (Fabaceae) (hummingbird and passerine bird flowers), and *Inga* (Fabaceae) (diurnal and nocturnal states of the nectar) by Baker and Baker (1982a, b, c). There is agreement with the generalizations. However, in *Mutisia viciaefolia* (Asteraceae) there is hexose-rich nectar in a species with a strongly hummingbird-adapted syndrome of characters. This last case is viewed as one of phylogenetic constraint in view of the general conservatism of the Asteraceae in nectar sugars (Baker and Baker 1982a).

Hybrids between species within the genera *Erythrina*, *Penstemon*, and *Campsis* show intermediate conditions in the F_1 generation (with greater resemblance to the "sucrose-dominant" parent in *Penstemon* and to the hexose-dominant parent in *Campsis* (Baker and Baker 1982a, b, c). Thus, it appears that the genetics of sugar ratios is complex.

Obviously, there are many evolutionary questions that remain to be answered regarding the relations between nectar-sugar chemistry and pollinator behavior, and it may be expected that this subject will receive increased attention comparable to that being given to energetics in nectar biology.

Nectar-Sugar Concentrations   The concentrations of sugars present in nectar (and the volume of the nectar itself) have received much attention, chiefly in terms of the energy requirements of the visitors utilizing this nectar (reviews by Heinrich 1975, 1979 cover this ground). In such calculations, however, one should be aware that pocket refractometer readings of "% sucrose equivalents" are on a "weight per total weight" basis which must be converted to "weight per volume" if caloric contents of nectars are to be estimated (Bolten et al. 1979). In addition, Inouye et al. (1980) have shown that lipids and amino acids in concentrations that may occur in nectars may have a small but occasionally significant effect on the refractive index of nectar and, therefore, on calculations based on this index.

Sugar concentrations in nectar may be affected by environmental conditions influencing nectar production (Percival 1965; Shuel 1955, 1957) and evaporation of secreted nectar (Fahn 1949; Corbet 1978a, b, c, etc.). Nectar in tubular flowers is less subject to this than that in open bowl-shaped flowers (F. L. Carpenter, pers. comm..). Baker (1978) reported that, for four out of five species of dry forest trees examined in Costa Rica, the sugar concentrations of samples of freshly produced nectar taken from different heights on the same tree (at the same time) differed slightly, with that of the upper flowers being greater. This may have some significance in relation to the patterns of bee-foraging at different heights observed by Frankie (1976).

In 1953, Hocking showed that the nectar of arctic tundra plants was particularly concentrated and this is also our experience in the alpine tundra of Colorado (Baker and Baker, unpub.). Also dry season concentrations in dry forest plants in Costa Rica are higher than those for cloud forests and wet lowland forests in Costa Rica (Baker 1975, 1978; Baker and Baker, unpub.).

Nevertheless, there is a general relation between nectar sugar concentration and pollinator type in any climatic zone (Baker 1975, 1978; also table 4.2). Recent papers on sugar concentrations include Brewer et al. (1974), Gut et al. (1977), Rust (1977), Stephenson and Thomas (1977), Whitham (1977), Corbet (1978a, b, c), Cruden and Hermann-Parker (1979), Wilson et al. (1979), Galen and Kevan (1980), Schemske (1980), and Sazima and Sazima (1980).

The concentrations of sugars in nectars that are taken up through narrow tubes by such visitors as Lepidoptera are lower (and consequently less viscous) than the nectars imbibed by bees and flies which have less narrow mouthparts. Also, hummingbirds, a portion of

TABLE 4.2
SUGAR CONCENTRATIONS OF NECTARS (ON A
SUCROSE-EQUIVALENT, WEIGHT PER TOTAL WEIGHT BASIS)
OF PLANTS OF VARIOUS POLLINATOR CLASSES IN LOWLAND
DRY FOREST, CLOUD FOREST, AND ALPINE AND SUBALPINE
ZONES

	Lowland Dry Forest (Costa Rica)		Cloud Forest (Costa Rica)		Alpine + Subalpine (Colorado)	
	N	$\bar{x}$(%)	N	$\bar{x}$(%)	N	$\bar{x}$(%)
Short-tongued bees + flies	5	46	19	21	41	32
Long-tongued bees	15	46	12	24	27	37
Bees + butterflies	–	–	8	22	26	27
Butterflies	20	29	7	14	4	24
Settling moths	2	41	8	18	1	18
Hawkmoths	11	24	8	15	–	–
Bats	5	17	4	15	–	–
Hummingbirds	29	21	20	17	3	41
	87		86		102	

Source: Baker and Baker 1982b.

whose tongue fills by capillarity (Baker 1975), and microchiropteran bats, who spend only a second or less at a flower (Baker 1978), select for relatively dilute, low viscosity nectar. While the concentrations of sugars in bee-pollinated species may reach as high as 88 percent (Kartashova 1965), it is unusual for the nectar of hummingbird flowers to exceed ca. 25 percent sucrose equivalent (weight by total weight). Bolten and Feinsinger (1978) have suggested that this might represent a "strategy" by the hummingbird flowers that avoids loss of their nectar to bee "thieves," but this seems unlikely in view of the small difference between the concentrations that they quote for flowers accessible to bees and those with the nectar held out of reach of bees' mouthparts ($\bar{x}$ = 21.3%, against $\bar{x}$ = 28.2%).

Many kinds of insects, including Diptera, Hymenoptera, and Lepidoptera can regurgitate liquid onto concentrated or even crystallized nectar and, in this way, reduce its concentration so that it may be imbibed.

AMINO ACIDS

The first demonstration of the presence of amino acids in detectable quantities in floral nectar appears to have been that by Ziegler (1956). Their presence in nectar as rewards to insects that have no other source of protein-building materials in the adult diet except nectar would be expected in view of the extraordinary range of food items tackled by various Lepidoptera, for example. Thus, some butterflies are known that take liquid sustenance from decaying flesh, faeces, urine, stagnant water, phloem sap (Ford 1945; Klots 1958) rotting fruit (Young 1972), and even the droppings of birds that follow army ants in the tropics (Ray and Andrews 1980). Female *Heliconius* butterflies are well known to collect pollen and "dunk" it in nectar to get an ingestible liquid with supplemented amino acid content (Gilbert 1972, 1975; Dunlop-Pianka et al. 1977), and it is possible that at least some species of butterflies in the genera *Battus* and *Parides* do the same (de Vries 1979). Moths are known that drink fruit juices, sweat, mucus from the eyes of animals, and even mammalian blood (Bänziger 1971). All of these are sources of organic nitrogen compounds and it is reasonable that floral nectars should be selected to play a similar role.

Analyses of amino acids in nectars published between 1956 and 1975 are reviewed by Baker and Baker (1975) and to the papers referred to there must be added a review by Kartashova (1965), and papers by Cruden and Toledo (1977), Rust (1977), Keeler (1977, 1980), Cruden and Hermann-Parker (1979), Gilliam et al. (1980), Scogin (1979a, b, 1980) and Voss et al. (1980).

Nectar Amino Acid Concentration Our methods of estimating amino acid concentrations by ninhydrin staining of nectar spots and comparison with the "histidine scale" are described in Baker and Baker (1973, 1975).

Although amino acid concentrations in nectars are considerably lower than sugar concentrations, they are not negligible. Calculations have shown (Baker and Baker 1973) that it takes no more than twenty flowers of *Dianthus barbatus* (Caryophyllaceae), a typical butterfly flower, to provide 840 nmol. of amino acids, an amount that, on a daily basis, has been shown by Gilbert (1972) to have a profound effect on the life span and reproductive output of *Heliconius* butterflies. Similarly, in the case of the orioles and tanagers (perching birds) that set up feeding territories in Mexico in flowering trees of *Erythrina breviflora* (Fabaceae), the range of individual amino acids and their

concentrations are enough to supply the needs of these birds for protein-building materials during the period of flowering of the trees
(Cruden and Toledo 1977; Cruden and Hermann-Parker 1977).

Even concentrations of amino acids that are too slight to be important nutritionally may contribute to the "taste" of the nectar, and
Kartashova (1965) has suggested that asparagine and glutamine, in
particular, are "anti-crystallizers" to sugars in nectar.

The concentration of amino acids in nectar, like that of the sugars,
is influenced by evaporation from open flowers (but, in our own studies, we have been careful to collect freshly produced nectar when concentrations were to be estimated). As with the sugar components of
nectar, there is a suggestion of correlation between the height of a
flower on a tropical tree and its nectar amino acid concentration but,
in this case, the correlation seems to be an inverse one (Baker 1978).
However, more than just environmental influences may be involved
here for, within a particular family, there is usually a higher mean
amino acid content for nectars from herbaceous plants compared to
those of woody plants (table 4.3).

The strongest amino acid concentrations in nectar are found in
flowers that simulate carrion or dung and attract the females of flies
that use these substrates as larval food (table 4.4). These flowers look
like and smell like the animal substances and when they produce nectar, it is appropriate that this should be rich in amino acids. By contrast, flowers visited by less specialized flies, in many cases, are also
visited by short-tongued bees and are generally only modestly supplied with amino acids. Long-tongued bees are also modestly supplied
with amino acids from the flowers they visit.

The high concentration of amino acids in flowers visited by carrion
and dung flies constitutes part of the "lure" of the female flies away
from their natural ovipositing sites. However, in general, the concentrations of nectar amino acids seem to be greater if nectar is the only
or the predominant source of protein-producing substances for the
flower visitor. Thus, the concentrations of amino acids in nectars in-
bibed by butterflies, settling moths, and wasps show high levels of
concentration. The nectars of most hummingbird and passerine bird
flowers and those collected by bats are weak in amino acids, and this
relates to the insect catching and pollen consumption, respectively,

TABLE 4.3
MEAN AMINO ACID CONCENTRATIONS FOR NECTAR
SAMPLES TAKEN FROM WOODY (TREES AND SHRUBS)
SPECIES AND HERBACEOUS SPECIES OF LARGE FAMILIES

	Amino Acids in Micromoles per ml.	
	Woody Species	*Herbaceous Species*
Apocynaceae	0.74	4.69
Araliaceae	0.25	1.02
Convolvulaceae	0.63	3.75
Crassulaceae	0.27	0.74
Fabaceae	0.63	1.02
Hydrophyllaceae	0.59	0.66
Lamiaceae	0.39	0.70
Onagraceae	0.37	0.47
Polemoniaceae	0.51	0.70
Rosaceae	0.37	0.47
Saxifragaceae	0.27	0.78
Scrophulariaceae	0.19	0.66
Solanaceae	0.31	0.59
Verbenaceae	0.33	0.37

Source: Baker and Baker (unpub.).

of the birds and the bats. Presumably, nectar could not supply adequate amounts for these rather large animals and there appears to have been no selection for strong amino acid content. An exception to the bird flower nectar picture is provided by the nectar of those *Erythrina* species (Fabaceae) which are pollinated by passerine birds (Baker and Baker 1982b, c). Thus, in the case of *E. breviflora* in Mexico, the orioles and tanagers that pollinate it set up feeding territories during the flowering period of the trees and are dependent upon nectar for their sustenance during this time (Cruden and Toledo 1977; Cruden and Hermann-Parker 1977).

Another apparent anomaly is provided by the hawkmoth flowers that generally produce nectar with relatively low amino acid contents (Baker and Baker 1975; Baker 1978), but it should be borne in mind that these insects consume large quantities of nectar each night (Baker and Baker 1975) and this means that the actual *amounts* of amino acids ingested may be considerable.

Data for individual taxa of angiosperms are provided in accounts

TABLE 4.4

MEAN AMINO ACID CONCENTRATIONS IN FLORAL
NECTARS, GROUPED ACCORDING TO PRINCIPAL TYPE
OF POLLINATOR

Principal Pollinators	Number of Determinations[a]	Amino Acids in Micromoles per ml.
Carrion & dung flies	9	12.500
Butterflies	118	1.500
Settling moths	78	1.059
Bees + Butterflies	257	1.015
Wasps	44	.913
Bees	715	.624
Flies (general)	97	.557
Hawkmoths	65	.536
Hummingbirds[b]	150	.452
Bats	23	.306
Passerine birds[c]	21	.255

Source: Baker and Baker 1982b.
[a]Determinations made in Berkeley from nectar collections in various tropical and temperate regions.
[b]Excludes *Erythrina* hummingbird-pollinated nectars.
[c]Excludes *Erythrina* passerine bird-pollinated nectars.

of the floral biology of *Gossypium* (Hanny and Elmore 1974), *Impatiens* (Rust 1977), *Erythrina* (Baker and Baker 1979b, 1982c), *Cheirostemon* and *Kigelia* (Scogin 1980), *Markea* (Voss et al. 1980).

Nectar Amino Acid Complements The early analyses of amino acid complements of nectars were carried out by paper chromatography (Ziegler 1956; Lüttge 1961, 1962; Maslowski and Mostowska 1963; Nair et al. 1964; Kartashova and Novikova 1964; Mostowska 1965; Baker and Baker 1973, 1975). Some estimate of the insensitivity of this method for amino acid analysis is provided by the small number of acids usually reported and the unlikely finding by Mostowska (1965) that in four taxonomically unrelated species there were identical complements of amino acids. This was shown to be erroneous (Baker and Baker 1977).

Our own analyses were made by paper chromatography in the earlier years (Baker and Baker 1973, 1975), but we have since adopted polyamide thin-layer chromatography of the dansylated amino acids with quantification (as well as identification) of individual acids by their fluorescence in U. V. light (Baker and Baker 1977, 1978). On a suitably miniaturized scale this provides rapid, repeatable, and relatively cheap results on amounts of nectar as little as 1–2 μl. The results show consistency in multiple determinations from separate plants, often under different environmental conditions (Baker and Baker 1977; Baker 1978). Other workers who have used automatic amino acid analysers and gas liquid chromatography have been able to detect amino acids at even lower concentrations but they need more nectar to make a "run" (e.g., Hanny and Elmore 1974; Gilliam et al. 1980; Rust 1977, all of whom have contributed amino acid complement data for individual species). Paper chromatography and TLC were used by Scogin (1980) in finding the complements of two bat pollinated species (*Cheirostemon platanoides* and *Kigelia pinnata*) and by Willmer (1981) for *Phaseolus, Mahonia,* and *Lamium.*

A summary of the proportions of nectars containing the various protein-building amino acids is given in table 4.5 (data from Baker and Baker 1982b). There are wide differences in the frequency of occurrence of the individual amino acids. There may be as few as two amino acids in a nectar (although many more than this are normal) up to about twenty-two acids (including some nonprotein amino acids to be discussed later). Nectar from *E. breviflora* was the first to show all ten "essential" amino acids in the same nectar (I. Baker in Cruden and Toledo 1977).

Inheritance of amino acid complements in crosses seems to be additive in the F_1 generation (on a presence or absence basis) (Baker and Baker 1976, 1977, 1979a, 1982b, c). Hybrid swarms in *Aquilegia* and *Penstemon* (Baker and Baker 1976) and in *Raphanus* (Baker and Baker 1977) show recombinations in the hybrid generations after the first and in backcrosses. The value of nectar amino acid complements (especially in view of additive inheritance) in resolving problems of allopolyploidy versus autopolyploidy and other phylogenetic concerns has been pointed out (Baker and Baker 1976, 1977, and unpub.)

The similarity (although not identity) between the nectar amino acid complements of related species and general immunity to environmental influences suggest strong genetic control of amino acid production. This, in turn, suggests the likelihood of a high degree of phylogenetic constraint. Thus, in the genus *Penstemon* (and the

TABLE 4.5
FREQUENCIES OF OCCURRENCE OF
INDIVIDUAL AMINO ACIDS IN FLORAL NECTARS
OF 395 SPECIES FROM VARIOUS TROPICAL AND
TEMPERATE REGIONS

	Detected In	Proportion
Alanine	380	.96
Arginine	356	.90
Serine	352	.89
Proline	344	.87
Glycine	332	.84
Isoleucine	287	.73
Threonine	263	.67
Valine	260	.66
Leucine	255	.66
Glutamic	245	.62
Cysteine, etc.	218	.55
Phenylalanine	216	.55
Tyrosine	204	.52
Tryptophan	189	.48
Lysine	162	.41
Glutamine	162	.41
Aspartic	128	.32
Asparagine	106	.27
Methionine	80	.20
Histidine	77	.19
Nonprotein	144	.36

Source: Baker and Baker 1982b.

closely related *Keckiella*) there is great similarity in the complements of members of the same subgenera, independent of the pollinator type (to which adaptation seems to be more a matter of concentration than identity of amino acids) (Baker and Baker 1981b). The same is true for investigated cases in *Erythrina* (Baker and Baker 1981b, c).

Nonprotein Amino Acids Many nectars that have been analyzed have contained amino acids other than the twenty conventionally denoted as protein-builders. Sometimes these nonprotein amino acids have been identified as γ-amino-butyric acid, β-alanine, ornithine, etc., but often they have had to be referred to as "unknown no. 1," etc. Of 248 floral nectars, 35 percent contained one or more nonprotein amino acids (Baker et al. 1978), and of these many contained more than one at the level of sensitivity of the TLC of dansylated acids (*Umbellularia californica* showed four) (Baker and Baker unpub.). Using an automatic analyzer, Rust (1977) found seven of these acids in nectar of *Impatiens capensis*.

Some nonprotein acids in seeds have been suggested to be a toxic defense against seed predators, for example, 3,4-dihydroxyphenylalanine (1-dopa) in *Mucuna* seeds (Rehr et al. 1973). They may perform a similar role in nectar (for they appear to be present more frequently in tropical than in temperate zone nectars (Baker 1977, 1978), but it is clear that they cannot be toxic to the pollinators. However, Rhoades and Bergdahl (1981) have postulated an alternative hypothesis wherein chemical defenses would be expected to characterize the rarer species in a flora (which would produce an exceptionally rich reward that will make it worthwhile searching for by specialized pollinators but also emphasizing their need for defense mechanisms against nectar robbers). This will be tested for in future studies.

It may be that some of these "non-protein" amino acids are not very toxic (e.g., γ-amino butyric acid) and they may function in some other capacity than as a chemical defense. When they can be identified they may be useful as phylogenetic guides, as are nonprotein amino acids in vegetative parts and seeds (Fowden 1970; Bell 1972).

OTHER POTENTIAL REWARDS

Lipids   The discovery of lipids as liquid rewards to pollinating insects visiting flowers was made by Stefan Vogel (1969, 1971, 1974) in South American species of two genera of Malpighiaceae, Krameriaceae, four genera of Scrophulariaceae (including *Calceolaria*), and four genera of Iridaceae (including *Sisyrinchium*) and Orchidaceae (*Oncidium*). He referred to the glands that produced these oils as "elaiophors" and considered them to be separate (and alternative) organs to nectaries. He observed that several genera of anthophorid bees, most obviously the genus *Centris* (*sensu lato*), collected these lipids from the flowers and, mixing them with pollen, used them as foodstuff for their larvae. Chemical analysis of the lipids from *Calceolaria pavonii* showed the major constituent to be a diglyceride of acetic acid and β-acetoxypalmitic acid (Vogel 1971, 1974). Simpson et al. (1977) examined the lipids from *Krameria* spp. and identified them as free saturated fatty acids with chain lengths between C16 and C20, with an acetate moiety at the β-position (and with no trace of glycerol). S. Buchmann (pers. comm.) has also described the secretion of lipids by glands on the connective of the stamens in *Mouriri* (Melastomataceae) and their manipulation by bee visitors to the flowers. In 1976, Vogel reported lipid production in flowers of *Lysimachia vulgaris* (Primulaceae) from Europe, where the oil is collected by bees of the genus *Macropis*.

However, in 1973, Baker and Baker reported for the first time that many *nectars* contain lipids (Baker and Baker 1973, 1975, 1982b; Baker 1977, 1978)—about 30 percent of those we have sampled. The test was a simple one (osmic acid staining of nectar spots on chromatography paper) and may not pick up all lipids, so a staining of fresh nectar drops with Sudan III was performed whenever possible. Lipid presence varies from a trace to quantities sufficient to make the nectar milky in appearance (e.g., *Catalpa speciosa* and *Jacaranda acutifolia*, Bignoniaceae, and *Trichocereus andalgalensis*, Cactaceae). There is a suggestion that in tropical forests in Costa Rica, lipids in nectar occur more frequently in trees than in any other life form. More detailed analyses are in progress to identify the constituents of these truly nectar lipids.

Antioxidants   Ascorbic acid (vitamin C) has been known to occur in honey since 1938 (Kask 1938) and in nectar since Griebel and Hess (1940) found it strongly represented in the nectar of three species of Lamiaceae. Weber (1942) identified it in *Fritillaria* nectar and he also found it in *Impatiens* (Weber 1951). Bukatsch and Wildner (1956) and Ziegler et al. (1964) added twelve more taxa. Our own studies (Baker and Baker 1975 and unpub.) have demonstrated the presence of organic "reducing" acids especially in many of the nectars that contain lipids, and it may be a function of the acids to prevent rancidity developing in standing nectar. Antioxidants tend to occur with lipids in stigmatic exudates (Baker et al. 1973) and they have been found in the oily elaiosomes of three species of flowering plant that have ant-distributed seeds (Bukatsch and Wildner 1956).

pH   Whatever may be the cause of the acidity or alkalinity, nectars display a wide range of pH (when freshly produced). *Silene alba* (Caryophyllaceae) nectar has shown a pH of 3 while, at the opposite extreme, *Viburnum costaricanum* (Caprifoliaceae) nectar reaches pH 10. The contributions of various chemicals (some of which may be inorganic) to this picture and the pollinatory significance of this range of pH will be the subject of future research.

Proteins It has been concluded (Baker and Baker 1975, 1982b) that the proteins that are often detectable in nectars are probably most frequently enzymatic. Beutler (1936) reported invertase in *Tilia* (Tiliaceae) floral nectar and Frey-Wyssling et al. (1954) and Matile (1956) found the same enzyme in other species. Loper et al. (1976) imply its presence in *Citrus* cultivars. Zimmerman (1953) found trans-glucosidase in *Robinia pseudoacacia* (Fabaceae) and Lüttge (1961) found tyrosinase in the nectar of *Lathraea* (Orobanchaceae). Phosphatases and oxidases have been recorded by Lüttge and Schnepf (1976), and ascorbic acid oxidase by Zauralov (1969). Recently, Scogin (1979) demonstrated the presence of esterases and malate dehydrogenase in nectar of *Fremontia* (Sterculiaceae). It is certain that many enzymes must occur in floral nectar and their identification and distribution must precede any conclusion as to their function.

UNFAVORABLE CHEMICALS

Alkaloids In addition to the substances whose presence in nectar can be looked upon as nutritionally positive to flower visitors, there are other substances that may have a deterrent or detrimental effect on certain visitors, presumably those that are nectar robbers rather than efficient cross-pollinators (but see also discussion under "nonprotein amino acids). The heterogeneous group of chemical substances classed as "alkaloids" may be examples.

Clinch et al. (1972) provided circumstantial evidence that alkaloids in the nectar of *Sophora microphylla* (Fabaceae) growing in New Zealand are severely detrimental to the health of honeybees. Deinzer et al. (1977) demonstrated the presence of pyrrolizidine alkaloids in the honey made from nectar of *Senecio jacobaea* (Asteraceae), although it is not clear whether the alkaloids were present in the nectar itself or in pollen contaminating it. The subject of alkaloids in nectar is discussed in Baker and Baker (1975: 127–129) along with a report of their presence in nectars of at least eight species of temperate zone plants. More cases have been demonstrated since then of the presence of "Dragendorff-positive" substances in floral nectar (Baker and

Baker 1982b). However, Baker (1977, 1978; Baker and Baker 1982b) provided statistics that demonstrate the possible presence of alkaloids in higher proportion in tropical forest plant nectars (12 percent) than in those from temperate zones (c. 7 percent). They are especially rare in the alpine zone of the Colorado Rockies (0 percent)—see Baker and Baker (1982b) for the latest figures on this subject, involving 850 species.

Of particular interest is the presence of alkaloids in the vegetative parts and stigmatic exudate of *Nicotiana sylvestris* (Solanaceae) but their apparent absence from the nectar produced by this moth-pollinated species (Baker et al. 1973).

It should be pointed out that there may be a rapid turnover of alkaloidal nitrogen in plants (Robinson 1968, 1974) so that the alkaloids may also function as temporary storage products. This would have significance for their presence in nectar if that solution were eventually resorbed, which does not usually take place (but see Kartashova 1965, who suggests that resorbtion of nectar occurs in many instances).

Phenolics Phenolic substances are another diverse group of chemicals which also may play a role in defense of nectar from thieves. Their presence in nectars from 333 out of 850 species color-tested with p-nitraniline (Baker 1977, 1978) suggests that they are widespread. Again, there appears to be a larger proportion of nectars giving a "positive" result in the tropical forests (49 percent) than in temperate California (30 percent) and the Colorado mountains (ca. 32 percent) (Baker and Baker 1982b). This group of chemicals needs further examination and closer identification before the possibly wide range of biological effects that they have can be appreciated. Only Scogin (1979a, b) has attempted an identification (as far as we are aware) in finding an isoflavone (5,7-dimethoxygenistein 4'-glucoside) in *Fremontia* (Sterculiaceae) nectar.

Miscellaneous Substances Among other potentially "unfavorable" chemicals in nectar may be listed inorganic substances. Thus, Waller (1973) showed that the attractiveness of the nectar of onion plants (*Allium cepa*, Alliaceae) to honeybees varied inversely with the amount of potassium present in the nectar.

Crane (1977, 1978) has reported that *Tilia* (Tiliaceae) nectar can be toxic to bees and Madel (1977) has shown that as few as two flowers produce enough toxin to kill a bumblebee (four species of *Bombus*) in a few hours. He has demonstrated the presence of the monosaccharide mannose in the nectar and ascribes the lethality of the nectar to this sugar.

Saponins are implicated in the toxicity of the nectar of *Aesculus hippocastanum* to honey bees by Schulz-Langner (1967). The toxic glycoside arbutin (hydroquinone plus glucose) has been isolated from honey derived from nectar of *Arbutus unedo* (Ericaceae) (Pryce-Jones 1944), and another glucoside was found in the nectar of *Ledum palustre* of the same family (Kozlova 1957), while Carey et al. (1959) obtained acetylandromedol from nectars of species of *Rhododendron* (also Ericaceae). Majak et al. (1980) found miserotoxin (3-nitro-1-propyl-β-D-glucopyranoside) in nectar of *Astragalus miser* var. *serotinus*, toxic to honey bees. Palmer-Jones (1968) found that nectar of the New Zealand tree *Corynocarpus laevigata* (Corynocarpaceae) is toxic to bees and the active substance, karakin, was identified by Finnegan and Stephani (1968) as 1,2,6-tri(3-nitropropanyl)-β-D-glucopyranose. Bell (1971) concluded that nectar of *Angelica triquetra* has a narcotic effect on bumblebees and other Hymenoptera. Other reports of toxic nectar can be found in Mauritzio (1945), Jaeger (1961), Palmer-Jones and Line (1962), and Stephenson (1981).

Other chemicals that have been found in nectar, but are probably of very restricted occurrence are listed in Baker and Baker (1975: 102), along with the appropriate references. Another list is Lüttge (1977). To these lists should be added "water soluble vitamins" (Ziegler et al. 1964). Clearly we have only begun to find and appreciate the biological significance of the wide range of chemicals that occur in nectar which, until recently, was generally considered nothing more than "sugar-water."

REFERENCES

(Papers by H. G. Baker and I. Baker are listed chronologically, irrespective of who is senior author)

Bachman, W. W. and G. D. Waller. 1977. Honeybee responses to sugar solutions of different compositions. *J. Apicultural Res.* 16:165–169.

Baker, H. G. 1975. Sugar concentrations in nectars from hummingbird flowers. *Biotropica* 7:37–41.

Baker, H. G. 1977. Non-sugar components of nectar. *Apidologie* 8:349–356.

Baker, H. G. 1978. Chemical aspects of the pollination biology of woody plants in the tropics. In P. B. Tomlinson and M. H. Zimmerman, eds., *Tropical Trees as Living Systems*, pp. 57–82. Cambridge & New York: Cambridge University Press.

Baker, H. G. and I. Baker. 1973. Some anthecological aspects of the evolution of nectar-producing flowers, particularly amino acid production in nectar. In V. H. Heywood, ed., *Taxonomy and Ecology*, pp. 243–264. London: Academic Press.

Baker, H. G. and I. Baker. 1975. Nectar constitution and pollinator-plant coevolution. In L. E. Gilbert and P. H. Raven, eds., *Animal and Plant Coevolution*, pp. 100–140. Austin: University of Texas Press.

Baker, I. and H. G. Baker. 1976. Analyses of amino acids in flower nectars of hybrids and their parents, with phylogenetic implications. *New Phytol.* 76:87–98.

Baker, H. G. and I. Baker. 1977. Intraspecific constancy of floral nectar amino acid complements. *Bot. Gaz.* 138:183–191.

Baker, H. G. and I. Baker. 1979a. Sugar ratios in nectars. *Phytochem. Bull.* 12:43–45.

Baker, I. and H. G. Baker. 1979b. Chemical constituents of the nectars of two *Erythrina* species and their hybrid. *Ann. Missouri Bot. Gard.* 66:446–450.

Baker, H. G. and I. Baker. 1982a. Floral nectar sugar constituents in relation to pollinator type. In C. E. Jones and R. J. Little, eds., *Handbook of Experimental Pollination Biology.* New York: Van Nostrand-Reinhold (in press).

Baker, H. G. and I. Baker. 1982b. Chemical constituents of nectar in relation to pollination mechanisms and phylogeny. In M. H. Nitecki, ed., *Biochemical Aspects of Evolutionary Biology*, pp. 131–171. Chicago: University of Chicago Press.

Baker, I. and H. G. Baker. 1982c. Some chemical constituents of floral nectars of *Erythrina* in relation to pollinators and systematics. *Allertonia* 3(1): 25–38.

Baker, H. G., I. Baker, and P. A. Opler. 1973. Stigmatic exudates and pollination. In N. B. M. Brantjes and H. F. Linskens, eds., *Pollination and Dispersal*, pp. 47–60. Nijmegen: University of Nijmegen.

Baker, H. G., P. A. Opler, and I. Baker. 1978. A comparison of the amino acid complements of floral and extrafloral nectars. *Bot. Gaz.* 139:322–332.

Banziger, H. 1981. Bloodsucking moths of Malaya. *Fauna* 1:4–16.

Baskin, S. I. and C. A. Bliss. 1969. Sugar occurring in the extrafloral exudates of the Orchidaceae. *Phytochemistry* 8:1139–1145.

Bell, C. R. 1971. Breeding systems and floral biology in the Umbelliferae. In

V. H. Heywood, ed., *The Biology and Chemistry of the Umbelliferae*, pp. 93–107. London: Academic Press.

Bell, E. A. 1972. Toxic amino acids in the Leguminosae. In J. B. Harborne, ed., *Phytochemical Ecology*, pp. 163–178. London: Academic Press.

Beutler, R. 1930. Biologisch-chemische Untersuchungen am Nektar von Immenblumen. *Zeit. Vergleich. Physiol.* 12:72–176.

Beutler, R. 1935. Nectar. *Bee World* 24:106–116, 128–136, 156–162.

Bolten, A. B. and P. Feinsinger. 1978. Why do hummingbird flowers secrete dilute nectar? *Biotropica* 10:307–309.

Bolten, A. B., P. Feinsinger, H. G. Baker, and I. Baker. 1979. On the calculation of sugar concentration in flower nectar. *Oecologia* (Berlin) 41:301–304.

Bonnier, G. 1878. Les nectaires. *Ann. Sci. Nat.* Serie 6, Bot. 8:5–212.

Brewer, J. W., K. J. Colyard, and C. E. Lott Jr. 1974. Analysis of sugars in dwarf mistletoe nectar. *Canad. J. Bot.* 52:2533–2538.

Bukatsch, F. and G. Wilder. 1956. Ascorbinsaürebestimmung in Nektar, Pollen, Blutenteilen, und Elaiosomen mit Hilfe einer neuen Mikromethode. *Phyton* 7:34–46.

Carey, F. M., J. I. Lewis, J. L. MacGregor, and M. Martin-Smith. 1959. Pharmacological and chemical observations on some toxic nectars. *J. Pharm. and Pharmacol.* Suppl. II, pp. 269T–274T.

Caspary, J. X. R. 1848. Dissertatio inauguralis de Nectariis. Bonn. (Quoted in Lorch 1978, q.v.).

Clinch, P. G., T. Palmer-Jones, and I. W. Forster. 1972. Effect on honeybees of nectar from the Yellow Kowhai (*Sophora microphylla* Ait.) *N.Z. J. Agric. Res.* 15:194–201.

Coon, C. S. 1955. *The History of Man*. London: Johnathon Cape.

Corbet, S. A. 1978a. Bees and the nectar of *Echium vulgare*. In A. J. Richards, ed., *The Pollination of Flowers by Insects*, pp. 21–29. London: Academic Press.

Corbet, S. A. 1978b. A bee's view of nectar. *Bee World* 59:25–32.

Corbet, S. A. 1978c. Bee visits and the nectar of *Echium vulgare* L. and *Sinapis alba* L. *Ecol. Entomol.* 3:25–37.

Crane, E. 1977. Dead bees under lime trees. *Bee World* 58:129–130.

Crane, E. 1978. Sugars poisonous to bees. *Bee World* 59:37–38.

Cruden, R. W. and S. Hermann-Parker. 1977. Defense of feeding sites by orioles and hepatic tanagers in Mexico. *The Auk* 94:594–596.

Cruden, R. W. and S. Hermann-Parker. 1979. Butterfly pollination of *Caesalpinia pulcherrima*, with observations on a psychophilous syndrome. *J. Ecol.* 67:155–169.

Cruden, R. W. and V. M. Toledo. 1977. Oriole pollination of *Erythrina breviflora* (Leguminosae): Evidence for a polytopic view of ornithophily. *Plant Syst. Evol.* 126:393–403.

Deinzer, M. L., P. A. Thomson, D. M. Burgett, and D. L. Isaacson. 1977.

Pyrrolizidine alkaloids: Their occurrence in honey from tansy ragwort (*Senecio jacobaea* L.). *Science* 195:497–499.

De Vries, P. J. 1979. Pollen-feeding rainforest *Parides* and *Battus* butterflies in Costa Rica. *Biotropica* 11:237–238.

Dunlop-Pianka, H., C. L. Boggs, and L. E. Gilbert. 1977. Ovarian dynamics in heliconiine butterflies: Programmed senescence versus eternal youth. *Science* 197:487–490.

Fahn, A. 1949. Studies in the ecology of nectar secretion. *Palest. J. Bot.* (Jerusalem) 4:207–224.

Finnegan, R. A. and R. A. Stephani. 1968. The structure of karakin. *Lloydia* 33:441.

Ford, E. B. 1956. *Butterflies*. London: Collins.

Fowden, L. 1970. The nonprotein amino acids of plants. *Progress in Phytochem.* 2:203–266.

Frankie, G. W. 1976. Pollination of widely dispersed trees by animals in Central America, with an emphasis on bee pollination systems: In J. Burley and B. T. Styles, eds., *Tropical Trees: Variation, Breeding, and Conservation*, pp. 151–159. London: Academic Press.

Frey-Wyssling, A., M. Zimmerman, and A. Mauritzio. 1954. Über den enzymatischen Zuckerumbau in Nektarien. *Experientia* 10:490–492.

Galen, C. and P. G. Kevan. 1980. Scent and color, floral polymorphisms, and pollination biology in *Polemonium viscosum* Nutt. *Amer. Midl. Nat.* 104: 281–289.

Gilbert, L. E. 1972. Pollen feeding and reproductive biology of *Heliconius* butterflies. *Proc. Nat. Acad. Sci.*, U.S.A. 69:1403–1407.

Gilbert, L. E. 1975. Ecological consequences of a coevolved mutualism between butterflies and plants. In L. E. Gilbert and P. H. Raven, eds., *Coevolution of Animals and Plants*, pp. 210–240. Austin: University of Texas Press.

Gilliam, M., W. F. McCaughey, and B. Wintermute. 1980. Amino acids in pollens and nectars of citrus cultivars and in bee breads and honeys from colonies of honey bees, *Apis mellifera*, placed in citrus groves. *J. Apicultural Res.* 19(1):64–72.

Gottsberger, G., J. Schrauwen, and H. F. Linskens. 1983. Die Zucker-bestandteil des Nektars einiger tropischer Blüten. *Portug. Acta Biol.* Series A, 13:1–8.

Grant, K. A. and V. Grant. 1968. *Hummingbirds and Their Flowers*. New York: Columbia University Press.

Griebel, C. and G. Hess. 1940. The vitamin C content of flower nectar of certain Labiatae. (In German: *Zeit. Untersuch Lebensmitt.* 79:168–171.) *Chem. Abstr.*, vol. 34; no. 44187 (1940).

Gut, L. J., R. A. Schlising, and C. E. Stopher. 1977. Nectar-sugar concentrations and flower visitors in the Western Great Basin. *Ecology* 37:523–529.

Hainsworth, F. R. and L. L. Wolf. 1976. Nectar characteristics and food selection by hummingbirds. *Oecologia* (Berlin) 25:101–113.

Hanny, B. W. and C. D. Elmore. 1974. Amino acid composition of cotton nectar. *J. Agric. and Food Chem.* 22:476–478.

Harborne, J. B. 1977. *Introduction to Ecological Biochemistry.* London: Academic Press.

Heinrich, B. 1975. The role of energetics in bumblebee flower interrelationships. In L. E. Gilbert and P. H. Raven, eds., *Coevolution of Animals and Plants,* pp. 141–158. Austin: University of Texas Press.

Heinrich, B. 1979. *Bumblebee Economics.* Cambridge: Harvard University Press.

Hocking, B. 1953. The intrinsic range and speed of flight of insects. *Trans. Royal Entomol. Soc.* (London) 104:223–345.

Inouye, D. W., N. D. Favre, J. A. Lanum, D. M. Levine, J. B. Myers, M. S. Roberts, F. C. Tsao, and Y. Y. Wang. 1980. The effects of nonsugar constituents on estimates of nectar energy content. *Ecology* 61:992–996.

Jeffrey, D. C., J. Arditti, and N. Koopowitz. 1970. Sugar content in floral and extrafloral exudates of orchids: Pollination, myrmecology, and chemotaxonomy implication. *New Phytol.* 69:187–195.

Kartashova, N. N. 1965. *Structure and Function of Nectaries of Dicotyledonous Flowering Plants.* (In Russian). Tomsk: Izdatel'stvo Tomskogo Universiteta.

Kartashova, N. N. and T. N. Novikova. 1964. A chromatographic study of the chemical composition of nectar. (In Russian: *Izv. Tomskogo Otd. Vses. Botan. Obshchestva* 5:111–119), *Chem. Abstr.* vol. 64, no. 8641a (1966).

Kask, M. 1938. Vitamin C — Gehalt der estnischen Hönigen. *Zeit. Untersuch Lebensmitt.* 76:543–545.

Keeler, K. H. 1977. The extrafloral nectaries of *Ipomoea carnea* (Convolvulaceae). *Amer. J. Bot.* 64:482–488.

Keeler, K. H. 1980. The extrafloral nectaries of *Ipomoea leptophylla. Amer. J. Bot.* 67:216–222.

Klots, A. B. 1958. *The World of Butterflies and Moths.* New York: McGraw-Hill.

Knuth, P. 1906–1909. *Handbook of Flower Pollination.* J. R. A. Davis, tr. Oxford: University Press.

Kozlova, M. V. 1957. Nectar of *Ledum palustre* as a possible source of the toxicity of honey. (In Polish: *Voprosy Pitaniya* 16:80.) *Chem. Abstr.,* vol. 52, no. 5695g (1958).

Loper, G. M., G. D. Waller, and R. L. Berdal. 1976. Effect of flower age on sucrose content of nectar in *Citrus. Hortscience* 11:416–417.

Lorch, J. 1978. The discovery of nectar and nectaries and its relation to views on flowers and insects. *Isis* 69:514–533.

Lüttge, U. 1961. Über die Zusammensetzung des Nektars und den Mechanismus seiner Sekretion. I. *Planta* (Berlin) 56:189–212.

Lüttge, U. 1962. Über die Zusammensetzung des Nektars und den Mechanismus seiner Sekretion. II and III. *Planta* (Berlin) 59:108–114, 175–194.

Lüttge, U. 1977. Nectar composition and membrane transport of sugars and amino acids: A review of the present state of nectar research. *Apidologie* 8:305–320.

Lüttge, U. and E. Schnepf. 1976. Elimination processes by glands: Organic substances. In *Encyclopedia of Plant Physiology*. New Series. *Transport in Plants*, vol. 2, part B, U. Lüttge and M. G. Pitman, eds., *Tissues and Organs*. Berlin: Springer.

Madel, G. 1977. Vergiftungen von Hummeln durch den Nektar der Silverlinde, *Tilia tomentosa* Moench. *Bonn zoologische Beitr.* 28:149–154.

Majak, W., R. Neufeld, and J. Corner. 1980. Toxicity of *Astragalus miser* var. *serotinus* to the honey bee. *J. Apicultural Res.* 19:196–199.

Maslowski, P. and I. Mostowska. 1963. Electrochromatographic estimation of free amino acids in honeys. (In Polish). *Pszcel. Zeszyty Nauk* 7:1–6 *Chem. Abstr.* vol. 60, no. 2258c (1964).

Matile, P. 1956. Über den Stoffwechsel und die Auxinabhangigkeit der Nektarsekretion. *Ber. Schweiz. Botan. Ges.* 66:237–296:

Mauritzio, A. 1945. Giftige Bienenpflanzen. *Beih. schweiz. Bienenztg.* 47:178–182.

Mauritzio, A. 1959. Papierchromatograpische untersuchungen an Blütenhonigen und Nektar. *Ann. de l'Abeille* 2:291–341.

Mostowska, I. 1965. Amino acids of nectars and honeys. (In Polish). *Zeszyty Nauk Wyzszej. Szkoly Rolniczej Olsztynie* 20:417–432. *Chem. Abstr.*, vol. 64, no. 20529 (1966).

Müller, H. 1883. *The Fertilisation of Flowers*. D'A. W. Thompson, ed. and tr. London: Macmillan.

Nair, A. G. R., S. Nagarajan, and S. Subramanian. 1964. Chemical compositions of nectar in *Thunbergia grandiflora*. *Current Sci.* (India) 33:401.

Niethammer, A. 1930. Mikrochemie einzelner Blüten in Zusammenhang mit der Honiggewinnung. *Gartenbauwiss.* 4:85–98.

Palmer-Jones, T. 1968. Nectar from karaka trees poisonous to honey bees. *N.Z. J.Agric.* 117:77.

Palmer-Jones, T. and L. J. S. Line. 1962. Poisoning of honey bees by nectar from the karaka tree. *N.Z. J. Agric. Res.* 5:433–436.

Percival, M. S. 1961. Types of nectar in angiosperms. *New Phytol.* 60:235–281.

Percival, M. S. 1965. *Floral Biology*. Oxford: Pergamon Press.

Pryce-Jones, J. 1944. Some problems associated with nectar, pollen, and honey. *Proc. Linn. Soc.* (London), pp. 129–174.

Ray, T. S. and C. C. Andrews. 1980. Antbutterflies: Butterflies that follow army ants to feed on antbird droppings. *Science* 210:1147–1148.

Rehr, S. S., D. H. Janzen, and P. P. Feeny. 1973. L-dopa in legume seeds: A chemical barrier to insect attack. *Science* 181:81–82.

Rhoades, D. F. and J. C. Bergdahl. 1981. Adaptive significance of toxic nectar. *Amer. Nat.* 117:798–803.

Robinson, T. 1968. *The Biochemistry of Alkaloids.* Berlin: Springer.

Robinson, T. 1974. Metabolism and function of alkaloids in plants. *Science* 184:430–435.

Rust, R. W. 1977. Pollination in *Impatiens capensis* and *Impatiens pallida* (Balsaminaceae). *Bull. Torrey Bot. Club* 104:361–367.

Sazima, M. and I. Sazima. 1980. Bat visits to *Marcgravia myriostigma* Tr. et Planch. (Marcgraviaceae) in southeastern Brazil. *Flora* 169:84–88.

Schemske, D. W. 1980. Floral ecology and hummingbird pollination of *Combretum farinosum* in Costa Rica. *Biotropica* 12:169–181.

Scogin, R. 1979a. Nectar constituents in the genus *Fremontia* (Sterculiaceae): Sugars, flavonoids and proteins. *Bot. Gaz.* 140:29–31.

Scogin, R. 1979b. 5,7-dimethoxy-4'-hydroxyisoflavone from *Fremontia* (Sterculiaceae). *Aliso* 9:479–480.

Scogin, R. 1980. Floral pigments and nectar constituents of two bat-pollinated plants: Coloration, nutritional, and energetic considerations. *Biotropica* 12:273–276.

Shuel, R. W. 1955. Nectar secretion. *Amer. Bee J.* 95:229–234.

Shuel, R. W. 1957. Some aspects of the relation between nectar secretion and nitrogen, phosphorus and potassium nutrition. *Canad. J. Plant Sci.* 37:220–236.

Simpson, B. B., J. L. Neff, and D. Seigler. 1977. Krameria, free fatty acids and oil-collecting bees. *Nature* 267:150–151.

Southwick, E. A. and A. K. Southwick. 1980. Energetics of feeding on tree sap by ruby-throated hummingbirds in Michigan. *Amer. Midl. Nat.* 104:328–334.

Stephenson, A. G. 1981. Toxic nectar deters nectar thieves of *Catalpa speciosa*. *Amer. Midl. Nat.* 105:381–383.

Stephenson, A. G. and W. W. Thomas. 1977. Diurnal and nocturnal pollination of *Catalpa speciosa* (Bignoniaceae). *Syst. Bot.* 2:191–198.

Stiles, F. G. 1976. Taste preferences, color preferences, and flower choice in hummingbirds. *Condor* 78:10–26.

Vogel, S. 1969. Flowers offering fatty oil instead of nectar. *Abstracts XIth Internatl. Bot. Congr.,* Seattle, p. 229.

Vogel, S. 1971. Oelproduzierende Blumen, die durch ölsammelnde Bienen bestaubt werden. *Naturwissenschaften* 58:58.

Vogel, S. 1974. Ölblumen und ölsammelnde Bienen. Akademie der Wissenschaften der Literatur Mainz: Tropische und Subtropische Pflanzenwelt 7:1–547.

Vogel, S. 1976. *Lysimachia:* Oelblumen der Holarktis. *Naturwissenschaften* 63:44–45.

Voss, R., M. Turner, R. Inouye, M. Fisher, and R. Cort. 1980. Floral biology of *Markea neurantha* Hemsley (Solanaceae), a bat-pollinated epiphyte. *Amer. Midl. Nat.* 103:262–268.

Waller, G. D. 1972. Evaluating responses of honey bees to sugar solutions using an artificial-flower feeder. *Ann. Entomol. Soc. Amer.* 65:857–862.

Waller, G. D. 1973. Chemical differences between nectar of onion and competing plant species and probable effects upon attractiveness to pollinators. Ph.D. dissertation, Utah State University.

Walker, A. K., D. K. Barnes, and B. Furgala. 1974. Genetic and environmental effects on quantity and quality of alfalfa nectar. *Crop Sci.* 14:235–238.

Watt, W. B., P. C. Hoch, and S. G. Mills. 1974. Nectar resource use by *Colias* butterflies. *Oecologia* (Berlin) 14:353–374.

Weber, F. 1942. Vitamin C in Nektar von *Fritillaria imperialis*. *Protoplasma* 36:316.

Weber, F. 1951. Impatiens-Nektar. *Phyton* (Horn) 3:110–111.

Whitham, T. G. 1977. Coevolution of foraging in *Bombus* and nectar dispensing in *Chilopsis:* A last dreg theory. *Science* 197:593–596.

Willmer, P. G. 1980. The effects of insect visitors on nectar constituents in temperate plants. *Oecologia* (Berlin) 47:270–277.

Willson, M. F., R. I. Bertin, and P. W. Price. 1979. Nectar production and flower visitors of *Asclepias verticillata*. *Amer. Midl. Nat.* 102:23–35.

Wykes, G. R. 1952a. The preferences of honey bees for solutions of various sugars which occur in nectar. *J. Exper. Biol.* 29:511–518.

Wykes, G. R. 1952b. The sugar content of nectars. *Biochem. J.* 53:294–296.

Young, A. M. 1972. Community ecology and some tropical rain forest butterflies. *Amer. Midl. Nat.* 87:146–157.

Zauralov, O. A. 1969. Oxidizing enzymes in nectaries and nectar. (In Russian.) *Trans. Nauch.-Issled Inst. Pchelovod.* 1969:197–225. *Chem. Abstr.*, vol. 74, no. 11050 (1971).

Ziegler, H. 1956. Untersuchungen über die Leitung und Sekretion der Assimilate. *Planta* (Berlin) 47:447–500.

Ziegler, H., U. Lüttge, and U. Lüttge. 1964. Die wasserlosichen Vitamine des Nektars. *Flora* (Jena) 154:215–229.

Zimmerman, M. 1953. Papierchromatographische Untersuchungen über die pflanzliche Zuckersekretion. *Ber. Schweiz Botan. Ges.* 63:402–429.

5

THE ECOLOGY OF NECTAR ROBBING

DAVID W. INOUYE
UNIVERSITY OF MARYLAND

It is generally accepted that flower forms represent mechanisms specialized for pollination by specific classes of pollinators (e.g., Grant and Grant 1965). The exclusion of other classes of pollinators is an important aspect of this specialization. One means of exclusion is the concealment of nectar at the base of a long corolla tube, which should circumvent the collection of nectar by visitors with short mouthparts. "But with regulation comes the possibility of evasion" (Proctor and Yeo 1973). There are two means of circumventing the barrier presented by a long corolla tube. A sympetalous (gamopetalous) corolla can only be avoided by perforating the corolla, or nectar robbing. A polypetalous corolla can be circumvented by forcing one's way between the individual sepals and petals forming the corolla tube.

Nectar robbing is a behavior exhibited by some species of birds, bees, and ants in which nectar is obtained through holes bitten near the bases of the corolla tubes, in a manner generally circumventing contact with the sexual parts of the flowers. The behavior occurs most frequently in situations in which the robber would not otherwise be able to reach the length of the corolla tube to extract nectar because of morphological constraints. Nectar robbers can be subdivided into primary nectar robbers, or those species or individuals which make the holes and then extract the nectar, and secondary robbers, or those species or individuals which obtain nectar by using holes made by primary robbers (Løken 1949). Secondary nectar robbers are generally unable to make the holes themselves. An additional category of flower visitors, the base workers (Weaver 1956), will also be considered in this discussion. These are the flower visitors which obtain nectar by working their way into the side of a polypetalous corolla.

Occasionally a flower visitor may, without biting or using holes, collect nectar from flowers morphologically adapted for pollination by a different class of visitors. To differentiate this from cases in which a hole is made or used, such behavior is referred to as nectar thieving (Inouye 1980b). In such cases nectar may be removed without contact between the sexual parts of the flower and the body of the visitor. For example, a butterfly might employ its long, thin proboscis to remove the nectar from a typically bumblebee pollinated flower and not contact the sexual parts of the flower which normally touch the bee's body. Although in some regards this would have the same effect on a flower as nectar robbing, such behavior will not be included in this discussion.

PRIMARY ROBBERS

The most notorious primary nectar robbers are short-tongued bees, in particular carpenter bees (*Xylocopa* sp.) and some species of bumblebees (*Bombus* sp.). New and Old World *Xylocopa* have been reported to rob flowers in at least 22 plant families (Barrows 1980). Bumblebees in both the New and Old World are also reported as primary robbers of a great variety of flowers. Lovell (1918) reports that at least 300 species of flowers are robbed. All of these bee species are of short proboscis length, and most occur in the subgenus *Bombus*. Most bumblebees, particularly those with long or intermediate proboscis lengths (Inouye 1976, 1977), are apparently psychologically or physically inept at making holes in flowers. The stingless bee *Trigona fulviventris* has also been reported as a primary robber (Barrows 1976). Although it is generally accepted that honeybees are physically unable to perforate corolla tubes (Free 1970), there are a few reports of *Apis* using their mandibles or even proboscides to perforate corollas (Gerner 1972).

The bees differ in the manner in which they make holes in the corollas. *Trigona fulviventris* is reported to bite holes in the corollas of

Lantana camara (Barrows 1976). *Xylocopa sonorina* has been observed to make perforations with its maxillae (Barrows 1977), a method probably used by all *Xylocopa* (van der Pijl 1954; Schremmer 1972). In Norway, *Bombus lucorum* and *sporadicus* use their proboscis to pierce the corolla tubes, while *B. mastrucatus* make quick cuts with their sharp, toothed mandibles (Løken 1949, 1962). In some instances, *B. lucorum* may use its maxillae, but *B. lapidarius* is said always to use its mandibles (Worth 1949, cited in Brian 1957). *B. terrestris*, closely related to *B. lucorum*, and *B. occidentalis* (in North America) also use their mandibles to make the holes (Gerner 1972; Eaton and Stewart 1969; Inouye, pers. obs.). There is some evidence that these species (and perhaps others commonly reported as nectar robbers) are morphologically adapted for their foraging behavior. Løken (1949) and Inouye (1976) found that the mandibles of *B. mastrucatus* and *B. occidentalis* are more toothed than those of bumblebee species not recorded as nectar robbers (fig. 5.1).

In a detailed photographic investigation of the foraging behavior of pollinators on *Aquilegia*, Macior (1966) found that in *Bombus affinis*, probing with the antennae, maxillae, and tongue was an important prelude to actual perforation by the mandibles. He suggested that the "sharp-pointed maxillae appear effective in locating nectar by slight perforation of the spur allowing the tip of the tongue to contact nectar . . . but it probably does not provide openings large enough for thorough nectar withdrawal. The more effective perforating mouthparts are undoubtedly the mandibles."

Other insects have also been implicated as nectar robbers. Rust (1977, 1979) has studied nectar robbing of *Impatiens capensis* by the

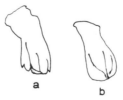

FIGURE 5.1
Mandibles of two species of bumblebees: *a. Bombus occidentalis*, a nectar-robbing species from western North America; *b. B. appositus*, a sympatric nonnectar-robbing species.

vespid wasp *Vespula maculifrons*. The floral tissue containing the nectary at the back of the jewel weed flowers was bitten through, or even off, by the yellowjackets. E. Stiles (pers. comm.) has observed nectar robbing by acantharid beetles on *Aureolaria pedicularia*.

Birds have also been reported as primary nectar robbers. The flower-piercers, the genus *Diglossa* in the honeycreeper family (Coerebidae), are perhaps the best-known examples. Skutch (1954) mentioned eleven species of flowers which were robbed by *Diglossa barritula*, the cinnamon-bellied flower piercer, and an additional flower was reported by Lyon and Chadek (1971). The slaty flower piercer (*D. plumbea*) competes with hummingbirds for the nectar of *Centropogon valerii* in Costa Rica (Colwell et al. 1974).

The manner in which the flower piercer punctured the flowers was described by Skutch (1954) and Colwell et al. (1974). The hooked upper mandible was placed over the flower to hold it in place "while the short upwardly tilted lower mandible pressed against the tissue and pierced through it" (Skutch 1954). The nectar could then be removed, presumably by inserting the tongue into the perforation.

Orioles are mentioned in one report as primary nectar robbers of *Ribes aureum* in Michigan (Darwin 1876:432). This role is consistent with other reports of nectar feeding by orioles (e.g., Schemske 1975). Hummingbirds are also reported occasionally to rob flowers. Grant (1952) found that hummingbirds would bite the spurs off of *Aquilegia pubescens*, a typically hawkmoth-pollinated flower, and Beal (1880) reported that hummingbirds visiting fuchsias would "pierce through the base of the calyx tube and take out the nectar" (see also Ingels 1976). Skutch (1954) stated that "an eccentric hummingbird will sometimes choose to pierce the base of a flower which others of its kind are visiting in the normal fashion." In these cases the bill was simply forced through the corolla. Janzen (1975) indicated that short-billed hummingbirds "often puncture the bases of long tubular flowers to get nectar, thereby making the flower less attractive to the long-billed hummingbird that normally pollinates the flower, and allowing bees to get at the nectar." While as Skutch indicates, this behavior is quite rare in general, primary robbing of tropical bignoneaceous flowers by hummingbirds is not uncommon (A. Gentry, pers. comm.; Gentry 1974).

Finally, ants have been recorded as primary nectar robbers, but apparently only in a single report (Wafa and Ibrahim 1959, cited in Free 1970). The activity of a mouse robbing nectar from *Fuchsia* flowers, observed on the University of California at Berkeley campus (H. Baker, pers. comm.), is probably highly unusual.

SECONDARY ROBBERS

A variety of bees have been observed as secondary robbers of floral nectar, using the holes provided by primary robbers. In a technical sense, a primary robber can also act as a secondary robber when it visits a flower in which it has already made a hole. Although there is one report of an *Osmia* species acting as a secondary robber (Gerner 1972), honeybees (*Apis mellifera*) and bumblebees are the most frequently observed secondary robbers. Moths have occasionally been observed using holes made by bees (pers. obs.), and in Costa Rica butterflies, particularly of the genera *Eurema* and *Phoebus*, are not uncommonly observed as secondary nectar robbers (P. Opler, pers. comm.). The abundance of secondary robbers on flowers is usually quite variable, reflecting in part their dependence on the activity of primary robbers in providing holes. Jany (1950) and Free (1968) both noted that the number of honeybees acting as secondary robbers on runner beans (*Phaseolus multiflorus*) was dependent on the number of robbing bumblebees.

BASE WORKERS

Although they do not do any damage to the flowers they visit, there are many similarities between base workers on polypetalous corollas, and nectar robbers on sympetalous corollas. There are no bee species which appear to specialize as base workers, but the behavior is reported most frequently from honeybees and those bumblebee species with short proboscides. Honeybees have been recorded as base workers on *Vicia villosa* (Weaver 1956, 1957, 1965), on Brussels sprout (*Brassica oleracea* cv *gemmifera*), bluebell (*Scilla nonscripta*), and wild cabbage (*Brassica oleracea*) (Free and Williams 1973). Bumblebees have been observed foraging as base workers on *Lathyrus odoratus* (Lovell 1918) and *L. leucanthus* (Inouye 1976). Ants occasionally collect nectar as base workers on wallflower (*Erysimum asperum*) (pers. obs.).

THE ONTOGENY OF NECTAR ROBBING

The importance of learned vs. instinctive components in the behavior of nectar robbers is disputed (Brian 1954, 1957; Macior 1966), as it is for many behaviors. It does seem clear, however, that a propensity toward nectar robbing was preceded by or has been accompanied with an increase in the flexibility of foraging behavior in certain species of bumblebees of short proboscis length. A reasonable hypothesis would seem to be that certain species of bees have a predisposition toward robbing flowers, but that the details of the behavior, such as where to bite flowers, must be learned. Observations by Kugler (1943) and Macior (1966) support this hypothesis. Kugler (1943) demonstrated that many parts of the flower are perforated at first, but once the bee has learned where the nectar is, it remembers this from flower to flower. Macior (1966) also suggested that the behavior of neonate workers indicated "that perforation behavior is innate but is not instinctively directed to nectar-bearing floral parts."

This hypothesis would explain the observations made by Herschel (1883) Kugler (1943), Løken (1949), Brian (1957), and Gerner (1972) that bilaterally symmetrical flowers have all been bitten on the same side in any one year. The bees probably make new holes in unperforated flowers on the same side in which they first succeeded in obtaining nectar. Observations on the behavior of left- and right-handed bees on *Lotus uliginosis* supported this conclusion (Brian 1957). It is not uncommon, however, to find bilateral flowers which have been perforated on both sides (pers. obs.).

Free and Williams (1973) suggested that it is because honeybees are "incompletely adapted to the flowers" that they will obtain nectar as base workers and as secondary robbers. A learning process similar to that involved in nectar robbing must be involved. The bees must discover that there is a means of access to the nectar other than that for which the flower has evolved. Weaver (1956) described the behavior of honeybees in the process of learning how to forage as base workers on hairy vetch (*Vicia villosa*). A bee can reach the nectar either "legitimately" by "tripping" the flower, or as a base worker without tripping the blossom, by inserting her proboscis between the petals of the standard and the keel at the base of the corolla tube. Individual bees with no previous experience on vetch flowers would "randomly attempt to insert the tongue into a blossom at any point. . . . Upon successfully working one blossom, or a very few blossoms in succession, most bees immediately became oriented in their behavior patterns, and calmly and methodically foraged from additional flowers by whatever method had proved successful." Free and William (1973) also studied honeybees as they learned to rob Brussels sprout. Their observations were in accordance with those of Weaver (1956). The bees "seemed to land anywhere on a flower or bud, and only gradually learned where the nectaries were located." Within a period of three to four days, most of the nectar-gathering bees were foraging as base workers.

The factors which lead one bee species to rob flowers regularly, while another species never does, are not resolved. As mentioned above, there is apparently a relationship between proboscis length of bumblebees and the appearance of the corolla perforating behavior. All the species commonly reported as nectar robbers are of short proboscis length, but not all species of short proboscis length are nectar robbers. There have been very few observations of bumblebees with very long proboscides acting as either primary or secondary nectar robbers. Brian (1957) investigated the possibility that the extended mouthparts of these species were relatively weak. She demonstrated that there was no significant difference in the strength of mandibles of queens of *B. hortorum* and *B. lucorum*, respectively, species of long and short proboscis length. She was not able to test the strength of maxillae, but suggested that difficulty in inserting a long proboscis into a small hole may be a more important factor. Kugler (1940) found that a long-proboscis bumblebee species "met with significantly more difficulty" than a short-proboscis species in feeding on honey droplets on a flat surface.

Given that there are apparently no major morphological restrictions preventing corolla biting in any bumblebee species, what other factors might be responsible for the observed behavioral differences? Brian (1957) suggested "that there are some grounds for supposing that an inherent difference in behaviour exists" between the common nectar robber and a nonnectar-robbing species that she studied.

She characterized the nonnectar robber as "conservative and stereo-typed, always visiting flowers by the obvious and correct entrance," and the nectar robber as "an opportunist obtaining nectar by a variety of methods." Other indications of the flexibility of feeding behavior of nectar-robbing species include the fact that they are generally the most unconventional in their methods of obtaining carbohydrates (Brian 1954). They collect honeydew (aphid secretions) from many plants (Brian 1954; Maurizio 1964), and may actually visit aphids in preference to flowers (Brian 1951). They have also been observed to visit ripe fruit (Hulkkonen 1928) and collect pollen from plants which exhibit features considered typical of the wind pollination syndrome (Pojar 1973). Richards (1975) calculated values for niche breadth of bumblebees visiting flowers in southern Alberta, and found that workers of *B. occidentalis*, commonly a nectar robber, had the larg-est value of the nine species examined for two consecutive years. Macior (1970, 1974) found that *B. occidentalis* had the greatest alti-tudinal range of the species he studied in Colorado.

EFFECTS ON POLLINATION

Many plants are dependent on insects for transfer of pollen from one flower to another, and insect visitation is generally equated with pollination. However, nectar robbers and base workers gener-ally by-pass the sexual parts of the flowers they visit, and primary nec-tar robbers cause some damage to the floral tissues. In some cases bumblebees may collect pollen from the same flowers which they rob, presumably thereby pollinating the flowers. Meidell (1944) and Koe-man-Kwak (1973) reported such behavior for bumblebees foraging from *Melampyrum pratense* and *Pedicularis palustris*, respectively. In both cases nectar robbing bumblebees collected pollen which was dusted on their legs and ventral surface after they vibrated their wings rapidly while in contact with the flower. This means of collecting pol-len is typically used on many flowers which produce pollen but no nec-tar reward for pollinators (e.g., Macior 1968, 1970), as well as those which also produce nectar. Some of these pollen foragers on *Melampy-rum* and *Pedicularis* alternated nectar robbing with pollen collecting, while other individuals either collected only nectar or pollen.

Although Soper (1952, cited in Alford 1975) has suggested that nectar robbers may effect pollination in self-fertile flowers simply by virtue of their movements on the flower, there is little evidence to suggest a beneficial effect on seed set from nectar robbing. Kendall and Smith (1976) examined pollination in the self-fertile runner bean (*Phaseolus coccineus*). Although honeybees and bumblebees that only robbed flowers produced significantly fewer pods than did bees foraging legitimately, there was a slightly higher seed set in robbed flowers than in control flowers, suggesting the effect which Soper postulated. Macior (1966) also suggested that nectar robbers often contact essential flower parts when landing on *Aquilegia* flowers. N. Waser (pers. comm.) found that Ocotillo (*Fouquieria splendens*) flowers robbed by *Xylocopa* had higher seed sets than flowers not visited by any pollinator or robber. He attributed this to the fact that the bees encountered the sexual parts of the flowers between landing on a flower and positioning themselves for robbing.

There are also several reports which indicate that nectar robbing has no harmful effect on seed set. For example, Brandenburg (1961) discovered that the sexual parts of broad bean flowers were not damaged by nectar robbers. This is probably generally true. The flowers which are most likely to be robbed are those with long corolla tubes, which have a long distance between the nectaries and the sexual parts of the flowers. For instance, tetraploid clover, which has longer corolla tubes than diploid clover, has a higher incidence of robbing than diploid clover (Friden et al. 1962, Valle et al. 1960). Billinski (1970) found that robbed flowers of *Vicia villosa*, *V. sativa*, *V. faba* and *Melampyrum pratense* had seed sets from 25 percent to 100 percent, and suggested that robbing probably has no harmful effect on these species.

Finally, there is at least one report that nectar robbing may result, although not directly, in an increased seed set. Hawkins (1961) found that in one of three years, there was a significant positive correlation between the activity of nectar robbers and seed set in red clover. He postulated that this was supporting evidence for the hypothesis proposed by Free and Butler (1959), that by making nectar accessible to honeybees seed set in a crop might be enhanced. The honeybees attracted to the crop might also collect pollen, thereby effecting pollination. However, as Kendall and Smith (1976) suggested, and as Free (1968) and Blackwall (1971) found, the activity of nectar robbers may induce honeybees which would otherwise act as legitimate foragers to forage instead as secondary nectar robbers. An alternative explanation, postulated by Heinrich and Raven (1972), is that "the actual pollinators (long-tongued bumblebees) visited more flowers when less nectar remained per flower," presumably to maintain some required level of nectar intake.

There is only one report documenting the effect of base workers on pollination. Free and Williams (1973) found that Brussels sprout flowers were less likely to be pollinated by base workers than legitimate foragers, and "the tendency of robbers to visit more flowers per plant and more plants per row than those that enter the flower decreases cross-pollination."

PROTECTION AGAINST NECTAR ROBBING

Given that there is evidence in some cases to suggest that nectar robbing can have a detrimental effect on seed set, it is not surprising to find that some flowers possess features which can be interpreted as offering protection against nectar robbers. These include both morphological and physiological adaptations.

It is generally accepted, if not well documented, that the function of extrafloral nectar is to attract ants for the purpose of protecting a plant. Faegri and van der Pijl (1966) cite an example in which the ant guard thus attracted keeps potential nectar robbers (*Xylocopa*) away from the base of the corolla of *Thunbergia grandiflora*. Flowers which are not successful in attracting an ant guard are subject to nectar robbing, but such flowers "hardly ever occur in nature" (Faegri and van der Pijl 1966). A similar system of ant protection of flowers from nectar robbers has been described for *Campsis radicans*, the trumpet creeper (Elias and Gelband 1975).

Structural adaptations are probably a relatively inexpensive means of protection against nectar robbing. Thickening of floral tissue, also found in *Thunbergia grandiflora*, is one potential means of discouraging nectar robbers. Reinforcement of the calyx and or bracts is another alternative. The butterfly flowers of the Pinks (*Dianthus*) are described by Proctor and Yeo (1973) as possessing "a firm leathery calyx, further protected at its base by stout overlapping bracts." They suggest that "the inflated calyces of the Campions (*Silene*) may serve the same function in a different way." In the flora of the Colorado Rocky Mountains, corolla tubes of *Lonicera involucrata* and species of *Castilleja*, both of the general floral morphology often robbed by *Bombus occidentalis*, appear to be protected by the presence of sufficient layers of tissue surrounding the corolla to discourage nectar robbing.

Another relatively simple means of protection may be the production of dense inflorescences instead of isolated flowers. Evidence of the effectiveness of such a structure is presented by Fogg (1950), who found that isolated flowers of Charlock (*Sinapsis arvensis*) were often robbed by base workers, while flowers in the denser inflorescences were rarely robbed. Proctor and Yeo (1973) suggest "that protection against nectar theft may have been one factor in the development of dense inflorescences seen, for instance, in Clovers (*Trifolium*, Leguminosae), *Buddleia* (Loganiaceae), Mints and Thyme (*Mentha* and *Thymus*, Labiatae), the Valerians (Valerianaceae), the Teasel and Scabious family (Dipsacaceae), and the Daisy family (Compositae)."

As a final possibility, the production of chemical deterrents to nectar robbing in certain flowers can be mentioned. Chemical deterrents to herbivory are widespread in plants, and it seems reasonable that they might be employed as a defense against nectar robbers. Because this is likely to be a relatively expensive means of deterring robbers, it should probably occur in relatively long-lived flowers with substantial quantities of nectar (P. Opler, pers. comm.). When injured, flowers of some tropical species of Apocynaceae produce a sap which is likely to be a chemical deterrent to nectar robbing (pers. obs.; P. Opler, pers. comm.).

WHY ROB FLOWERS?

One obvious answer to this question is that nectar robbing may be the only means by which a nectar feeder can get nectar from a particular flower species, because of a morphological mismatch. A second important answer was suggested over a hundred years ago by Charles Darwin (1876): "The motive which impels bees to gnaw holes through the corolla seems to be the saving of time, for they lost much time in climbing into and out of large flowers, and in forcing their heads into closed ones." In short, nectar robbing is potentially more efficient than legitimate flower visitation.

Several lines of evidence suggest that this is indeed the case for base workers. In a comparison of Brussels sprout cultivars, Free and Williams (1973) found positive correlations between the tendency of honeybees to rob and the depth and width of corolla tubes, the length and width of petals, and the length of filaments. These results seem to imply that the bees found it more difficult to obtain nectar by entering larger flowers, and turned to foraging as base workers. The increased rate of flower visitation by base workers observed by Free and Williams (1973) was interpreted as reflecting "the ease and hence the greater economy with which they collected nectar." The tendency of base workers to visit more flowers per plant and more plants per row than legitimate foragers is also consistent with the hypothesis that base workers collect nectar more efficiently.

Weaver (1956, 1957, 1965) presents data demonstrating the greater efficiency of base workers, compared to legitimate foragers, on hairy vetch (*Vicia villosa*). Legitimate foragers required a mean of 10.5 sec to visit a single flower, visiting a mean of 3.1 flowers per minute. Base workers were, depending on the period considered, 21 to 50 percent faster, taking a mean of 8.5 seconds per flower, and visiting a mean of 4.8 flowers per minute. Inouye (1976, 1980a), in an investigation of the relationship between proboscis length of bumblebees and how quickly they could visit flowers, recorded data for base workers and legitimate foragers on *Lathyrus leucanthus*. Predicting from the calculated least-squares regression equation for legitimate foragers (fig. 5.2), *Bombus bifarius* would take 6.6 seconds per flower as a legitimate forager. Foraging as a base worker, however, only required a mean of 4.6 seconds per flower. Another potential benefit of foraging as a base worker is indicated by Weaver (1956), who observed that "base workers did not appear to expend nearly so much muscular energy in reaching the nectar" as did legitimate foragers.

Similarly, it has been demonstrated that secondary nectar robbers can forage more efficiently than legitimate foragers on some flowers. Free (1968) found that honeybees could visit runner beans (*Phaseolus multiflorus*) 70 percent faster as secondary robbers than as legitimate foragers. Honeybees and bumblebees foraging on an ornamental

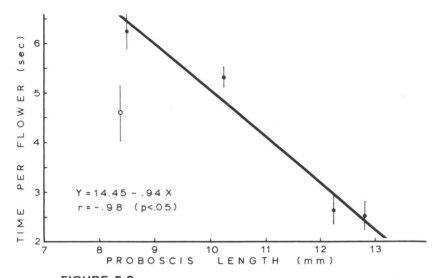

FIGURE 5.2

Data for handling times of bumblebee queens foraging legitimately on *Lathyrus leucanthus* (Leguminosae) in Colorado. Each point represents a mean value, verticle lines indicate ± one S.E. of the mean. The open circle, not included in the regression line, represents data for a queen of *Bombus bifarius* foraging as a base worker.

shrub, *Abelia grandiflora*, demonstrated a similar economy, visiting flowers 29 percent and 32 percent faster as secondary nectar robbers (Inouye, unpub. data). D. Morse (pers. comm.) recorded similar results from bumblebees foraging on cow vetch. Primary nectar robbers foraged significantly faster than individuals of the same species observed foraging legitimately, and a secondary robber foraged faster than other individuals of the same species that were foraging legitimately. These data probably provide an explanation for Weaver's (1965) observation that honeybees "spent a large percentage of their time searching for blossoms that had been punctured" previously by *Xylocopa*.

A primary nectar robber arriving at a patch of flowers is potentially faced with two different types of resource and a decision to make: is it more efficient to fly around and look for flowers with holes already in them, and ignore those without holes, or is it more rewarding to make holes in those flowers encountered without holes? At least four factors are likely to affect the decision, if the robber is capable of deciding on the most efficient behavior. First, the degree to which foraging as a secondary nectar robber is faster than making a hole is an important consideration. If it took no longer to make a hole than it did to use one there should be no reason to ignore flowers without holes. Second, the probability of encountering flowers with holes in them should affect the decision. If there is no probability of encountering flowers with holes, as when the first primary nectar robber arrives at a patch of flowers, the decision should be an easy one. Third, the rate at which the resource comprising flowers with holes is renewed is important. If, once a flower is visited, the rate at which nectar is secreted is negligible, then the robber should ignore flowers with holes. A closely related parameter is the amount of nectar likely to be found in flowers without holes, or more important, the probability of receiving a greater reward from an unopened flower than a cut one.

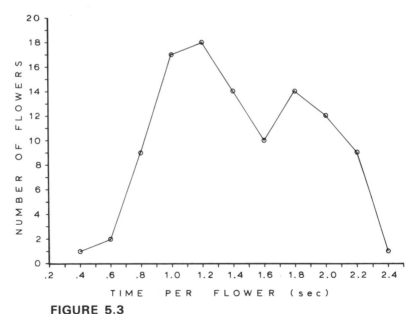

FIGURE 5.3

Distribution of times recorded for nectar-robbing bumblebees on *Abelia grandiflora*, Chapel Hill, N.C. The bimodal distribution reflects the differences in efficiencies of primary and secondary nectar robbers.

Relating to the first factor discussed, the relative efficiency of making and using holes, bumblebees can forage 65 percent faster on flowers of *Aconitum columbianum* if a hole is present than if they must make one. (Mean time 12.6 sec without a hole, 4.4 sec with a hole, difference significant, t-test, $p < .001$) (Inouye, unpub. data). If data for nectar robbers include both primary and secondary robbing, we might then expect to find a bimodal distribution of times, with one peak representing secondary robbers and the other primary robbers. Such a curve is presented in fig. 5.3 for bumblebees foraging on *Abelia grandiflora*.

No data are available in the literature pertaining to the remaining factors which should influence the robbers' decision to ignore flowers without holes instead of making holes. It seems a reasonable hypothesis, however, that if bumblebees are to maximize their intake of nectar, they should switch their behavior at some threshold from making holes to not making holes.

EFFECT OF NECTAR ROBBING ON SPECIES DIVERSITY OF BUMBLEBEE COMMUNITIES

Coexisting bumblebee species appear to partition flower resources primarily on the basis of the relationship between proboscis length and the length of the corolla tubes of the flowers available (Inouye 1977, 1980a). There are three categories of proboscis length in both Old and New World bumblebees: short, intermediate, and long. With one exception it appears to be generally true that each category can only be represented by one species in a group of coexisting bumblebee species. With the same exception, coexisting bumblebee species differ in mean proboscis length by a constant factor of 1.2 to 1.4 (Inouye 1977). The significance of this figure is suggested by the empirical observation that in certain groups of animals, otherwise similar species differ in the size of their feeding apparatus by a constant factor of 1.2 to 1.4 (Hutchinson 1959; Schoener 1974). This observation, as well as subsequent theoretical analyses (e.g., May 1974) suggest that there is a limit to how similar two species can be and still coexist. The exception to the pattern of size differences mentioned above is the coexistence of nectar robbing bumblebee species with other species of short proboscis length which are not nectar robbers.

Through the behavior of nectar robbing, bumblebees of short proboscis length are able to collect nectar from flowers they would otherwise be excluded from, including flowers which are not visited by any
other bumblebee species. For example, *Bombus occidentalis* often
nectar robs scarlet gilia (*Ipomopsis aggregata*), a flower which exhibits all the characters of the hummingbird pollination syndrome. (The
flowers are long, red, tubular, and contain comparatively copious
quantities of nectar). Thus a nectar-robbing bumblebee does not actually compete solely with other bee species of short proboscis length
for the same floral resources. Rather, it can utilize the whole contin-
uum of flower sizes, including flowers which are not available to
other bee species. Wratt (1968) reported that workers of *Bombus
terrestris* (a nectar robber) had versatile foraging habits and "inci-
dentally [!] avoided most competition from other species" by their
pattern of resource utilization. Interestingly, there are apparently no

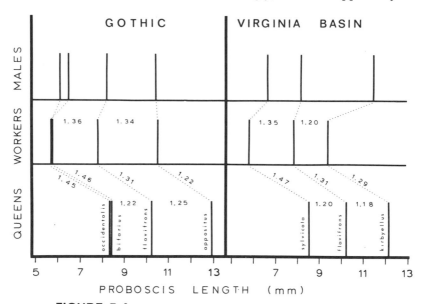

FIGURE 5.4

The distribution of proboscis lengths represented by coexisting species
of bumblebees at two study sites in Colorado. *Bombus occidentalis*, a
nectar-robbing species, overlaps strongly with *B. bifarius*, a nonnectar-
robbing species, in Gothic (elevation 2886 m) but is not present in Virginia Basin (elevation 3505 m).

broad areas in North America where more than one species of nectar robbing bumblebee is common (L. Macior, pers. comm.).

In Colorado, *Bombus occidentalis*, commonly a nectar robber, coexists with *B. bifarius* (not a nectar robber) in Gothic, at an elevation of 2886 m (9470 ft) (fig. 5.4) (Inouye 1976, 1977). Both species are of the same proboscis length, but *B. occidentalis* takes nectar from a much wider range of corolla tube lengths than does *B. bifarius*. Although *B. occidentalis* is not present in Virginia Basin, at an elevation of 3505 m (11,500 ft), there is still a bumblebee species of short proboscis length present, *B. sylvicola* (fig. 5.4). A pattern similar to that found in Colorado has also been documented from the Old World. Brian (1957), in a study of bumblebees in the area of the Firth of Clyde in Scotland, found only one common species representing intermediate and long proboscis lengths, but two coexisting species of short proboscis length, one of which (*B. lucorum*) is a nectar robber (fig. 5.5). It appears, therefore, that the behavior of nectar robbing

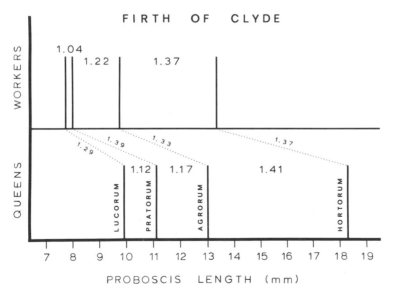

FIGURE 5.5

The distribution of proboscis lengths represented by coexisting species of bumblebees at the Firth of Clyde, Scotland. *Bombus lucorum* is a nectar-robbing species. (Data from Brian 1957).

in bumblebees permits the coexistence of a greater diversity of species than would otherwise be possible.

In summary, nectar robbing is a complex phenomenon. It involves a variety of organisms, and may have profound consequences for a variety of aspects of plant/flower visitor interactions. These include the competitive interactions among and species diversity of bumblebees, the morphology of flowers, and other aspects of the coevolutionary relationships between plants and flower visitors. Further studies of nectar robbing, in the context of the current interest in optimal foraging theory (e.g., Pyke, Pulliam, and Charnov 1977), should provide a more solid understanding of the behavior, at least from the robbers' point of view.

REFERENCES

Alford, D. V. 1975. *Bumblebees.* London: Davis-Poynter.

Barrows, E. M. 1976. Nectar robbing and pollination of *Lantana camera* (Verbenaceae). *Biotropica* 8:132–135.

Barrows, E. M. 1977. Floral maturation and insect visitors of *Pachyptera hymenaea* (Bignoniaceae). *Biotropica* 9:133–134.

Barrows, E. M. 1980. Robbing of exotic plants by introduced carpenter and honeybees in Hawaii, with comparative notes. *Biotropica* 12:23–29.

Beal, W. J. 1880. Fertilization of flowers by hummingbirds. *Amer. Nat.* 14: 126–127.

Bilinski, M. 1970. Results of observations on perforating flowers by bumblebees *Bombus* Latr. (Hym. Apoidea). (In Polish). *Polskie Pismo Entomol.* 40:122.

Blackwall, F. L. C. 1971. A study of the plant/insect relationships and podsetting in the runner bean (*Phaseolus multiflorus*). *J. Hort. Sci.* 46:365–379.

Brandenburg, W. 1961. Broad beans: causes of poor yields sought. *N.Z. J. Agric.* 102:277–280.

Brian, A. D. 1951. The pollen collected by bumble-bees. *J. Anim. Ecol.* 20: 191–194.

Brian, A. D. 1954. The Foraging of Bumble Bees. II. Bumble bees as pollinators. *Bee World* 35:81–91.

Brian, A. D. 1957. Differences in the flowers visited by four species of bumblebees and their causes. *J. Anim. Ecol.* 26:71–98.

Colwell, R. K., B. J. Betts, P. Bunnell, F. L. Carpenter, and P. Feinsinger. 1974. Competition for the nectar of *Centropogon valerii* by the hummingbird *Colibri thalassinus* and the flower-piercer *Diglossa plumbea*, and its evolutionary implications. *Condor* 76:447–452.

Darwin, C. 1876. *The Effects of Cross and Self-Fertilisation in the Vegetable Kingdom*. New York: Appleton.

Eaton, G. W. and M. G. Stewart. 1969. Blueberry damage caused by bumblebees. *Canad. Entomol.* 101:149–150.

Elias, T. S. and H. Gelband. 1975. Nectar: Its production and functions in trumpet creeper. *Science* 189:289–291.

Faegri, K. and L. van der Pijl. 1966. *The Principles of Pollination Ecology*. Oxford: Pergammon.

Fogg, G. E. 1950. Biological flora of the British Isles. *Sinapsis arvensis* L. *J. Ecol.* 38:415–429.

Free, J. B. 1968. The behavior of bees visiting runner beans (*Phaseolus multiflorus*). *J. Appl. Ecol.* 5:631–638.

Free, J. B. 1970. *Insect Pollination of Crops*. London: Academic Press.

Free, J. B. and C. G. Butler. 1959. *Bumblebees*. London: Collins.

Free, J. B. and I. H. Williams. 1973. The foraging behaviour of honeybees (*Apis mellifera* L.) on Brussels sprout (*Brassica oleracea* L.). *J. Appl. Ecol.* 10:489–499.

Friden, F., L. Eskillson and S. Bingefors. 1962. Bumblebees and red clover pollination in central Sweden. *Sveriges Froodl. Forb. medd.* 1:17–26.

Gentry, A. H. 1974. Coevolutionary patterns in Central American Bignoniaceae. *Ann. Missouri Bot. Garden* 61:728–759.

Gerner, W. 1972. Blüteneinbruch durch Apiden. *Zool. Anz.* (Leipzig) 189: 34–44.

Grant, V. 1952. Isolation and hybridization between *Aquilegia formosa* and *A. pubescens*. *Aliso* 2:341–360.

Grant, V. and K. A. Grant. 1965. *Flower Pollination in the Phlox Family*. New York: Columbia University Press.

Hawkins, R. P. 1961. Observations on the pollination of red clover by bees. I. The yield of seed in relation to the numbers and kinds of pollinators. *Ann. Appl. Biol.* 49:55–65.

Heinrich, B. and P. H. Raven. 1972. Energetics and pollination ecology. *Science* 187:597–602.

Herschel, I. 1883. Some habits of bees and bumble bees. *Nature* (London) 29: 104.

Hulkkonen, O. 1928. Zur Biologie der sudfinnischen Hummeln. *Ann. Univ. Aabo*. Series A, 3:1–81.

Hutchinson, G. E. 1959. Concluding remarks. *Cold Spring Harbor Symp. Quant. Biol.* 22:415–427.

Ingels, J. 1976. Observations on the hummingbirds *Orthorhyncus cristatus* and *Eulampis jugularis* of Martinique (West Indies). *Gerfaut* 66:129–132.

Inouye, D. W. 1976. Resource partitioning and community structure: A study of bumblebees in the Colorado Rocky Mountains. Ph.D. dissertation, University of North Carolina.

Inouye, D. W. 1977. Species structure of bumblebee communities in North

America and Europe. In W. J. Mattson, ed., *The Role of Arthropods in Forest Ecosystems*, pp. 35–40. New York: Springer.

Inouye, D. W. 1980a. The effect of proboscis and corolla tube lengths on patterns and rates of flower visitation by bumblebees. *Oecologia* (Berlin) 45: 197–201.

Inouye, D. W. 1980b. The terminology of floral larceny. *Ecology* 61:1251–1253.

Jany, E. 1950. Der 'Einbruch' von Erdhummeln (*Bombus terrestris* L.) in die Beuken der Feuerbohne (*Phaseolus multiflorus* Willd). *Zeit. angew. Entomol.* 32:172–183.

Janzen, D. H. 1975. *Ecology of Plants in the Tropics*. London: Edward Arnold.

Kendall, D. A. and B. D. Smith. 1976. The pollinating efficiency of honeybee and bumblebee visits to flowers of the runner bean *Phaseolus coccineus* L. *J. Appl. Ecol.* 13:749–752.

Koeman-Kwak, M. 1973. The pollination of *Pedicularis palustris* by nectar thieves (short-tongued bumblebees). *Acta Bot. Neerl.* 22:608–615.

Kugler, H. 1940. Die Bestaubung von Blumen durch Furchenbienen (*Halictus* Latr.). *Planta* 30:780–799.

Kugler, H. 1943. Hummeln als Blutenbesucher. *Ergebn. Biol.* 19:143–323.

Løken, A. 1949. Bumble bees in relation to *Aconitum septentrionale* in central Norway. *Nytt Magasin for naturvidenskapene* 87:1–60.

Løken, A. 1962. Occurrence and foraging behaviour of bumblebee visiting *Trifolium pratense* L. in Norway. *Meddelande Sveriges Fröodlareforbund* 7:52–59.

Lovell, J. H. 1918. *The Flower and the Bee*. New York: Scribner's.

Lyon, D. L. and C. Chadek. 1971. Exploitation of nectar resources by hummingbirds, bees (*Bombus*), and *Diglossa baritula* and its role in the evolution of *Penstemon kunthii*. *Condor* 73:246–248.

Macior, L. W. 1966. Foraging behavior of *Bombus* (Hymenoptera: Apidae) in relation to *Aquilegia* pollination. *Amer. J. Bot.* 53:302–309.

Macior, L. W. 1968. Pollination adaptation in *Pedicularis groenlandica*. *Amer. J. Bot.* 55:927–932.

Macior, L. W. 1970. Pollination ecology of *Dodecatheon amethystinum* (Primulacea). *Bull. Torrey Bot. Club* 97:150–153.

Macior, L. W. 1970. The pollination ecology of *Pedicularis* in Colorado. *Amer. J. Bot.* 57:716–728.

Macior, L. W. 1974. Pollination ecology of the front range of the Colorado Rocky Mountains. *Melanderia* 15.

Maurizio, A. 1964. Mikroskopische und Papierchromatographische Untersuchungen an Honig von Hummeln, Meliponinen und Anderen, zuckerhaltige Safte sammelnden Insekten. *Zeit. fur Bienenforschung* 7:98–110.

May, R. M. 1974. *Stability and Complexity in Model Ecosystems*. Princeton, N.J.: Princeton University Press.

Meidell, O. 1944. Notes on the pollination of *Melampyrum pratense* and the "honey stealing" of bumblebees and bees. *Bergens Mus. Arb.* 11:5–11.

Pijl, van der, L. 1954. *Xylocopa* and Flowers of the Tropics. I. The bees as pollinators: Lists of the flowers visited. *Proc. Koninkl. Nederl. Akad. van Wetenschappen* (Amsterdam). Series C, 57:413–423.

Pojar, J. 1973. Pollination of typically anemophilous salt marsh plants by bumble bees, *Bombus terricola occidentalis. Am. Midl. Nat.* 89:448–451.

Pyke, G. H., H. R. Pulliam, and E. L. Charnov. 1977. Optimal foraging: a selective review of theory and tests. *Q. Rev. Biol.* 52:137–154.

Proctor, M. and P. Yeo. 1973. *The Pollination of Flowers.* London: Collins.

Richards, K. W. 1975. Population ecology of bumblebees in southern Alberta. Ph.D. dissertation, University of Kansas.

Rust, R. W. 1977. Pollination in *Impatiens capensis* and *Impatiens pallida* (Balsaminaceae). *Bull. Torrey Bot. Club* 104:361–367.

Rust, R. W. 1979. Pollination of *Impatiens capensis:* Pollinators and nectar robbers. *J. Kansas Entomol. Soc.* 52:297–308.

Schemske, D. W. 1975. Territoriality in a nectar feeding Northern Oriole in Costa Rica. *Auk* 92:594–595.

Schoener, T. W. 1974. Resource partitioning in ecological communities. *Science* 185:27–39.

Schremmer, F. 1972. Der Stechsaugrussel, der Nektarraub, das Pollensammeln und der Blutenbesuch der Holzbienen (*Xylocopa*) (Hymenoptera, Apidae). *Z. morph. Tiere* 72:263–294.

Skutch, A. F. 1954. *Life Histories of Central American Birds.* Vol. 1. Pacific Coast Avifauna Series, no. 31. Berkeley, Calif.

Valle, O., M. Salminen, and E. Huokuna. 1960. Pollination and seed setting in tetraploid red clover in Finland II. *Acta Agralia Fennica* 97:1–63.

Weaver, N. 1956. The Foraging Behavior of Honeybees on Hairy Vetch. I. Foraging methods and learning to forage. *Insects Sociaux* 3:547–549.

Weaver, N. 1957. The Foraging Behaviour of Honeybees on Hairy Vetch. II. The foraging area and foraging speed. *Insects Sociaux* 4:43–57.

Weaver, N. 1965. The Foraging Behaviour of Honeybees on Hairy Vetch. III. Differences in the vetch. *Insects Sociaux* 12:321–326.

Wratt, E. C. 1968. The pollinating activities of bumble bees and honeybees in relation to temperature, competing forage plants, and competition from other foragers. *J. Apic. Res.* 7:61–66.

6

EXTRAFLORAL NECTARIES: THEIR STRUCTURE AND DISTRIBUTION

THOMAS S. ELIAS
NEW YORK BOTANICAL GARDEN
CARY ARBORETUM

Nectaries are organs or specialized tissue that secrete a substance known as nectar, composed of monosaccharides and disaccharides, amino acids, proteins, and trace amounts of other compounds. Nectaries and nectar are intimately linked with many vital functions of flowering plants, the most commonly known being the events leading to pollination. Extrafloral nectaries are important in maintaining the mutually beneficial relationship between many plants and certain insects, especially ants, which are attracted to the nectaries and in turn offer the plant varying degrees of antiherbivore protection. Also, the presence of ants on extrafloral nectary sites on outer floral parts may deter other organisms that reduce reproductive capacity of plants by avoiding normal pollination procedures and robbing nectar.

Extrafloral nectaries can be distinguished from floral nectaries by either position (topography) or function. Caspary (1848) distinguished between the two on the basis of position, that is, nectaries on any floral part were "floral," while nectaries on vegetative structures were "extrafloral." Delpino's (1875) system of defining nectaries accorded to function, floral nectaries being involved with pollination whereas extrafloral ones are involved with nonpollination functions.

The difference between the two systems is well illustrated by the two species of *Campsis* (Bignoniaceae) (Elias and Gelband 1975, 1976). Clusters of nectaries found on the upper surface of the petioles of young leaves, on calyx lobes, and on corolla lobes attract ants. Another large nectary surrounding the base of the ovary is intimately involved with pollination. A fifth site is the surface of young developing fruits. Those nectaries growing on a reproductive structure are neither evident nor functional until after fertilization when the corolla has fallen away and the fruits begin to develop, at which time they draw ant foragers. Thus, the nectaries on the calyx, corolla, and fruit are floral according to Caspary and extrafloral according to Delpino.

Extrafloral nectaries may occur on virtually all vegetative and reproductive structures. On leaves, they occur on the petiole, rachis when applicable, dorsal and ventral surfaces of the blade, and even the leaf margin. Nectaries are more likely to occur on the upper half of the petiole at or near the base of the leaf blade than at any other site. In some members of the Leguminosae and Rosaceae for example, either the stipules subtending the leaves contain nectaries or the entire stipular structure is nectariferous. Less common is the presence of nectaries on young stems, especially in the nodal region. Extrafloral nectaries have even been reported on cotyledons (Zimmerman 1932).

Nectaries associated with inflorescences and flowers, but not directly involved with the pollination process, are common. They may be scattered along the inflorescence rachis and appear as tiny bumps or tubercles near the base of the flowers or on their pedicels. Nectaries on or associated with bracts and involucres are commonly seen in Bignoniaceae and Malvaceae. They are present on calyx lobes and to a lesser extent on the corolla, generally the abaxial side.

Following pollination and fertilization, the ovary with or without other floral parts begins its development into a fruit. In several species of Bignoniaceae, nectaries are numerous on the surfaces of developing fruits. These are almost microscopic glands which attract ants with their nectar (Elias and Prance 1978). Nectaries are expected to occur on developing fruits in other, primarily tropical families.

Because of the diversity and wide distribution of extrafloral nectaries, the many avenues for their development, and their relationships to most other organs on the plant body, any attempt to provide a workable classification must be viewed with compassion. The most thorough survey of extrafloral nectaries in angiosperms is that of Zimmerman (1932); he developed a classification scheme based first on structure, then on topography. Another, specifically tropical study, conducted by Schell, Cusset, and Quenum (1963), consists of a family by family description of the nectaries encountered. Many other workers have contributed detailed information about nectaries in specific families, genera, or species (Arbo 1972; Bentley 1977; Bhattacharyya & Maheshwari 1971a, 1971b; Elias, Rozich, and Newcombe 1975; Elias and Gelband 1975, 1976; Ianishevskii 1941; Janda 1931; Siebert 1948; and Waddle 1970.)

Zimmerman's work is an often overlooked source of valuable information and his classification system remains useful. The basic structural and positional categories are: Gestaltlosennektarien, Flachnektarien, Grubennektarien, Hohlnektarien, Schuppennektarien, and Hochnektarien.

GESTALTLOSENNEKTARIEN (formless nectaries): These are amorphous nectaries that lack obvious structural specialization at the tissue or organ level but are capable of secreting a rich nectar. They can be recognized on the living plant by the presence of nectar and often a distinct coloring of the site of secretion. Nectar-producing sites on the floral bracts of *Paeonia* (fig. 6.1) and *Costus* are examples.

FLACHNEKTARIEN (flattened nectaries): These nectaries are flattened and closely pressed against the fundamental tissue of other organs so that the glandular surface is scarcely above or just beneath the surface level of the surrounding tissue (fig. 6.2). They are usually rounded or oval with a flattened, concave, or convex surface, and are common on the lower leaf surface of some species of *Luffa* (Cucurbitaceae), *Dioscorea* (Dioscoreaceae), and *Malpighia* (Malpighiaceae).

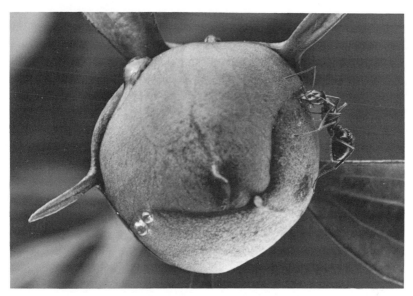

FIGURE 6.1
Flower bud of *Paeonia sp.* with nectar droplets secreted by the amorphous nectaries at the edges of the bud scales.

GRUBBENEKTARIEN (pit nectaries): These nectaries are sunken in the tissue of other organs. Depressions in which they lie are usually steep-sided and have a span whose diameter equals or exceeds that of the nectaries. Typical examples are found on the leaves of some species of *Ligustrum* (Oleaceae), *Gossypium* (fig. 6.3) and *Hibiscus* (Malvaceae), *Grewia* (Tiliaceae), and *Polygonum* (Polygonaceae).

HOHLNEKTARIEN (hollow nectaries): These are cavities in other organs with a narrow channel, slit, or pore extending to the surface. The cavities are usually lined with glandular-secreting trichomes. Zimmerman cites *Fagraea* (Loganiaceae), *Ipomoea* (Convolvulaceae), and *Hibiscus* (Malvaceae) among others as examples.

SCHUPPENNEKTARIEN (scalelike nectaries): These are specialized trichomes modified for nectar production and secretion. Most are scalelike, although some are cup-shaped and always occur as separate, distinct units. Both are often visible to the naked eye. Excellent examples are found on the floral bracts in *Thunbergia grandi-*

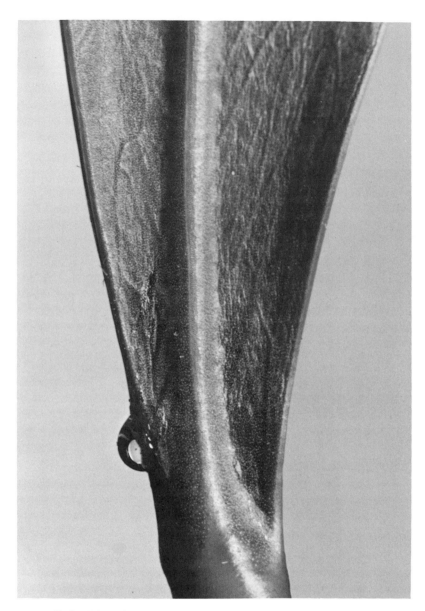

FIGURE 6.2
Side view of phyllode base of *Acacia longifolia* showing nectar droplet at site of flattened nectary.

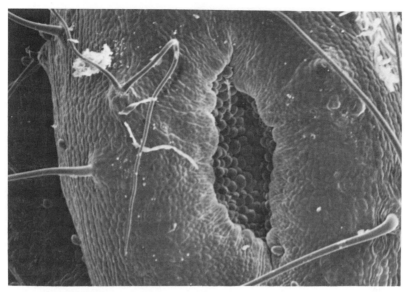

FIGURE 6.3
Nectary of *Gossypium hirsutum* embedded in the midvein on the lower leaf surface. The round-topped cells lining the cavity comprise the secretory tissues.

flora Roxb. (Acanthaceae), on the petiole and calyx of *Clerodendron* species (Verbenaceae), and on the petioles, leaves, flowers, and some fruits of most genera of Bignoniaceae (fig. 6.4).

HOCHNEKTARIEN (elevated nectaries): This category includes all nectaries, excepting the Schuppennektarien, which are distinctly raised above the ground tissue. It is a very common type of nectary, as seen on the leaves in *Hoya* species (Asclepiadaceae), lower leaf surface in *Terminalia paniculata* W. & A. (Combretaceae), base of the leaf blades in *Turnera* (Turneraceae), some species of *Prunus* (Rosaceae), and in many genera of Leguminosae (figs. 6.5, 6.6, 6.7).

Zimmerman further separated from the Hochnektarien those nectaries that supposedly came about through transformation of another organ. Examples include the flower bud nectaries in the leaf axils of some species of *Capparis* (Capparaceae), leaf tooth nectaries in *Impatiens* (Balsaminaceae), and "thorn nectaries" on some Cactaceae.

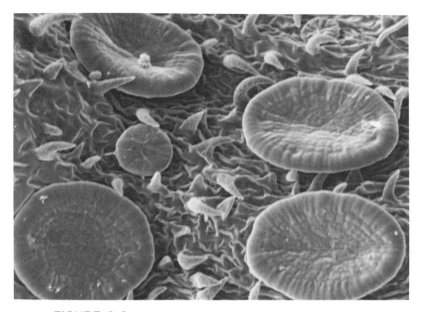

FIGURE 6.4
Four flattened, scalelike nectaries in the nodal position on the stem of *Bignonia capreolata*, a vine of the southern United States.

I have identified a new position in extrafloral nectaries that does not fit any of Zimmerman's categories (Elias, 1980). Therefore, I propose an additional type:

EINBETTENNEKTARIEN (embedded nectaries):   This type of nectary is totally embedded in tissues of other organs. Secretory cells of *Leonardoxa africana* form a disc-shaped gland enclosed in the leaf mesophyll. A tiny flasklike neck through which the nectar is secreted extends to the lower leaf surface.

STRUCTURE

Anatomically, extrafloral and floral nectaries are similar, albeit extrafloral ones are often larger and more varied in external appearance. Several independent lines of extrafloral nectary specialization have evolved; some can be identified with certain families or orders as evidenced by table 6.1. These evolutionary lines include nonvascularized nectaries without a well-defined structure, nonvascular-

FIGURE 6.5

Pinnate leaf of *Inga spuria* showing immature leaflets, winged rachis, and an elevated, cup-shaped nectary between the insertion of each pair of pinnae.

FIGURE 6.6

Close up of the nectary of *Inga spuria* with nectar droplet.

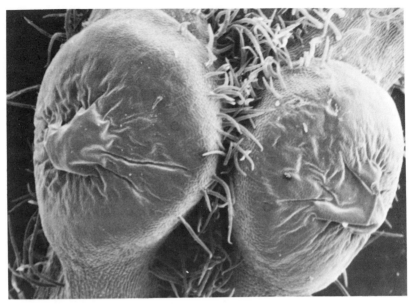

FIGURE 6.7
Paired nectaries of *Passiflora incarnata* at the junction of the lamina and
the petiole. Nectar, prior to release, is stored under the central bulge.

ized nectaries with an organized structure, and the larger, vascularized nectaries.

   Nonvascularized Nectaries   There are several distinct modes
of nonvascularized nectaries. One of the most fascinating yet poorly
understood is the nectar-secreting zone on floral bracts of *Paeonia*.
Similar functional sites are present on floral bracts of *Costus* and
some members of the Cactaceae. They lack a defined structure, both
internally and externally, identifiable as a nectary. The only obvious
anatomical evidence of function in *Paeonia* is a separation of the
epidermis from the underlying cells. This apparently is due to up-
lifting by the accumulating nectar prior to its release.

   Pit nectaries can be seen in cotton, *Gossypium hirsutum* L. They
are located on the midvein near the base of the lower leaf surface
(Wergin et al. 1975). The nectaries appear as depressions in the
midvein. They are composed primarily of parenchyma cells and a

single row of papillate secretory cells. No vascular tissue is present in the subglandular tissues.

The most common type of nonvascularized extrafloral nectaries are scale-like structures. Those found throughout the primarily tropical Bignoniaceae are the most thoroughly studied of this type (Dop 1927; Elias and Gelband 1975, 1976; Inamdar 1969; Laroche 1974; Parija and Samal 1936; Rao 1926 and Seibert 1948).

Extrafloral nectaries in this family are small, apparently specialized trichomes which originate from papillate epidermal cells. Both nonglandular and glandular trichomes are also present on most species. The nectar-secreting trichomes seem to have evolved from glandular trichomes which in turn derived from nonglandular ones. Often the nectaries are clustered at predictable sites on the plant body. Two variations in shape are recognized: shallowly dish-shaped with a low rim (patelliform), and cup-shaped with an upturned rim (cupular). Internal structures are similar.

The secretory cells in both cupular and patelliform nectaries are narrow, columnar, and anticlinally arranged. There are one, or less commonly, two strata of secretory cells having very conspicuous nuclei and dense, darkly staining cytoplasm. A thin cuticle of overlapping, waxlike deposits covers the secretory cells and any other external part of the nectary.

A single basal cell or a row of basal cells beneath the secretory cells connect them with the epidermis. A single basal cell is present in all species of Catawba tree (*Catalpa*) and the tropical calabash trees (*Crescentia*). In *Campsis*, on the other hand, secretory cells are subtended by a single layer of cuboidal cells which have conspicuous nuclei and lighter staining cytoplasm (Elias and Gelband 1976).

The presence or absence of vascular tissue in the nectary is not necessarily an indication of evolutionary advancement, even though Frey-Wyssling (1955) proposed that the development of an organized structure supplied with vascular tissue is more specialized than one lacking it. Carlquist (1969) concluded that the amount of vascular tissue in a structure is directly proportional to its size and is not necessarily related to any state of advancement. While this theory is accepted, certain exceptions are conceded: in the Leguminosae, for example, the specialization of foliar nectaries into large, elaborately vascularized, cupular or stipitate structures does indicate specialization trends. Specialized types of nectaries in the Leguminosae correlate with other advanced characters such as pollen or polyad types.

Vascularized Nectaries   Unlike some of the nonvascularized nectaries, all known vascularized ones have a well-defined structure. These are variable in shape and size, ranging from sessile, trough, or saucerlike nectaries to raised stalked to cupular glands.

Vascularized extrafloral nectaries are larger than nonvascularized ones. The composition and extent of branching of the vascular supply is apparently dependent upon the size and shape of the nectaries. Frey-Wyssling (1955) distinguished three types of vascularization: exclusively by phloem, by phloem and xylem, and predominantly by xylem. Accordingly, the concentration of sugars in the nectar is greater in nectaries supplied exclusively by phloem than by xylem.

For example, many species of *Impatiens* (Balsaminaceae) have foliar nectaries at marginal teeth apices near the base of the lamina and sometimes extending onto the petiole (Elias and Gelband, 1977). The vascular tissue extending into this structurally simple nectary is composed of phloem with only a few strands of xylem. This is in contrast to the hydathodes, also located at the apices of the teeth, but more distal. They are vascularized solely with xylic elements.

Most nectaries are vascularized by both phloem and xylem. Commonly, the vascular strands terminate a few cells beneath the secretory cells, although phloic cells have been observed in contact with those cells. In my studies of extrafloral nectaries, I have not observed any that are vascularized solely or predominantly by xylem elements, as reported by Frey-Wyssling to occur in perigones of *Fritillaria imperialis*.

In the mimosoid legumes studied, the vascular supply to nectaries on the petiole and rachis are derived from secondary adaxial traces rather than the primary vascular bundle. The large, cupular nectaries found in *Inga* and *Pithecellobium* contain a much branched vascular system. The presence of such a system is considered a function of the size of nectaries and not an indication of a higher level of specialization.

Stipitate and large cupular nectaries, as seen in some Leguminosae, usually have a sclerenchymatous bundle sheath which encloses part or all of the vascular bundles, with fibers varying from one to several layers in thickness. They are especially evident in the large foliar nectaries of *Pithecellobium macradenium* (Elias 1972) (fig. 6.8).

Ground parenchyma cells associated with nectaries, particularly those close to the secretory cells, often contain greater deposits of

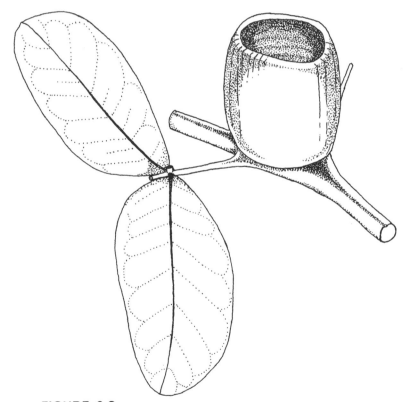

FIGURE 6.8
Leaf base of *Pithecellobium macradenium* showing large "gigas-type" nectary at the junction of the lowest pair of pinnae and a smaller, cupular nectary between the pair of leaflets.

tannins than parenchyma in adjoining organs. Furthermore, calcium oxalate raphides or druses in the ground parenchyma of nectaries are of widespread occurrence.

The secretory cells are located under the epidermis and may be one to several layers thick. These cells are most often elongated and oriented along a vertical axis. They have conspicuous nuclei and dense, darkly staining cytoplasm. Mitochondria are numerous and the endoplasmic reticulum abundant.

DISTRIBUTION

Presently, 68 families of flowering plants are known to have species which bear extrafloral nectaries, according to the functional definition of these nectaries. Additional families will likely be identified. This statistic is based not only on my own observations, but also on the reported observations of Zimmerman (1932) and Schell, Cusset, and Quenum (1963). As most of the earlier determinations were made upon gross structural elements, it is possible that a few of the taxa in the families listed here may have external glands other than nectaries.

In the six subclasses of dicotyledons, extrafloral nectaries are absent in the Magnoliales, scarce in Hamamelidae and become increasingly abundant in the more advanced subclasses Caryophyllidae, Dillenidae, Rosidae, and Asteridae. Extrafloral nectaries are especially prevalent in the latter three taxa.

On the whole, the monocotyledons (Class Liliatae) contain only a few members which have extrafloral nectaries. These are largely in the orders Zingiberales and Liliales; the Cyperales, Arales, and Orchidales are also known to contain species with nectaries.

The occurrence of these nectaries in 35 orders of angiosperms, together with their diverse ontogenies, is strong evidence that extrafloral nectaries have evolved independently many times in different families, and at different times during the evolutionary history of flowering plants. Although most extrafloral nectaries must be considered analogous, there are obvious homologies in related families in several orders. The shortstalked nectaries on the distal portion of the petiole in Turneraceae and Passifloraceae are good examples. Other possible examples can be seen in the Leguminosae/Rosaceae/

Chrysobalanaceae complex and in the Malvaceae/Bombacaceae/ Sterculiaceae complex of families.

Orders and Number of Families of Angiosperms with Extra-floral Nectaries All orders are listed with total number of families followed by the total number of families/families with extrafloral nectaries in parentheses; only those families known to contain species with extrafloral nectaries are listed. The phylogenetic sequence is based on Cronquist, *The Evolution and Classification of Flowering Plants*, 1968.

TABLE 6.1

Class MAGNOLIOPSIDA
(Dicotyledonaeae)
Subclass MAGNOLIIDAE
1. Order Magnoliales (19/0)
2. Order Piperales (3/0)
3. Order Aristolochiales (1/0)
4. Order Nymphales (3/0)
5. Order Ranunculales (8/0)
6. Order Papaverales (2/0)

Subclass HAMAMELIDAE
1. Order Trochodendrales (2/0)
2. Order Eucommiales (1/0)
3. Order Urticales (5/1)
   Moraceae—rare
4. Order Leitneriales (1/0)
5. Order Juglandales (3/0)
6. Order Myricales (1/0)
7. Order Fagales (3/0)
8. Order Casuarinales (1/0)

Subclass CARYOPHYLLIDAE
1. Order Caryophyllales (11/1)
   Cactaceae—common
2. Order Batales (1/0)
3. Order Polygonales (1/1)
   Polygonaceae—occasional

4. Order Plumbaginales (1/1)
   Plumbaginaceae—abundant

Subclass DILLENIIDAE
1. Order Dilleniales (3/1)
   Paeoniaceae—abundant
2. Order Theales (13/2)
   Dipterocarpaceae—rare
   Marcgraviaceae—common
3. Order Malvales (6/4)
   Tiliaceae—occasional
   Sterculiaceae—occasional
   Bombacaceae—occasional
   Malvaceae—common
4. Order Lecythidales (1/0)
5. Order Sarraceniales (3/2)
   Sarraceniaceae—common
   Nepenthaceae—common
6. Order Violales (21/6)
   Flacourtiaceae—rare
   Turneraceae—abundant
   Passifloraceae—abundant
   Bixaceae—common
   Dioncophyllaceae—common
   Curcurbitaceae—common

Class MAGNOLIOPSIDA
(Dicotyledonaeae) *cont.*
Subclass DILLENIIDAE *cont.*
7. Order Salicales (1/1)
   Salicaceae—occasional
8. Order Capparales (5/1)
   Capparaceae—rare
9. Order Ericales (9/1)
   Ericaceae—occasional
10. Order Diapensiales (1/0)
11. Order Ebenales (5/1)
    Ebenaceae—rare
12. Order Primulales (3/0)

Subclass ROSIDAE
1. Order Rosales (17/4)
   Hydrangeaceae—rare
   Rosaceae—common
   Chrysobalanaceae—occasional
   Leguminosae—abundant
2. Order Podostemales (1/0)
3. Order Haloragales (4/0)
4. Order Myrtales (13/2)
   Melastomataceae—rare
   Combretaceae—occasional
5. Order Proteales (2/0)
6. Order Cornales (2/0)
7. Order Santalales (10/0)
8. Order Rafflesiales (3/0)
9. Order Celastrales (10/1)
   Dichapetalaceae—occasional
10. Order Euphorbiales (5/1)
    Euphorbiaceae—common
11. Order Rhamnales (3/1)
    Rhamnaceae—rare
12. Order Sapindales (17/5)
    Sapindaceae—rare
    Anacardiaceae—occasional

Class MAGNOLIOPSIDA
(Dicotyledonaeae) *cont.*
Subclass ROSIDAE *cont.*
   Simaroubaceae—common
   Rutaceae—rare
   Meliaceae—rare
13. Order Geraniales (5/1)
    Balsaminaceae—abundant
14. Order Linales (3/2)
    Humiriaceae—rare
    Linaceae—rare
15. Order Polygalales (7/3)
    Malpighiaceae—common
    Vochysiaceae—common
    Polygalaceae—occasional
16. Order Umbrellales (2/0)

Subclass ASTERIDAE
1. Order Gentianales (4/3)
   Loganiaceae—rare
   Apocynaceae—rare
   Asclepiadaceae—occasional
2. Order Polemoniales (8/1)
   Convolvulaceae—occasional
3. Order Lamiales (1/1)
   Verbenaceae—occasional
4. Order Plantaginales (1/0)
5. Order Scrophulariales (12/5)
   Oleaceae—common
   Bignoniaceae—abundant
   Acanthaceae—common
   Pedaliaceae—occasional
   Scrophulariaceae—common
6. Order Campanulales (5/0)
7. Order Rubiales (1/0)
8. Order Dipsacales (5/1)
   Caprifoliaceae—rare
9. Order Asterales (1/1)
   Compositae—rare

Class LILIATAE
Subclass ALISMATIDAE
1. Order Triuridales (2/0)
2. Order Hydrocharitales (1/0)
3. Order Najadales (8/0)

Subclass COMMELINIDAE
1. Order Commelinales (4/0)
2. Order Eriocaulales (1/0)
3. Order Restionales (5/0)
4. Order Juncales (2/0)
5. Order Cyperales (2/1)
   Gramineae—rare
6. Order Typhales (2/0)
7. Order Bromeliales (1/0)
8. Order Zingiberales (8/4)
   Zingiberaceae—common
   Marantaceae—occasional

Class LILIATAE *cont.*
Subclass COMMELINIDAE *cont.*
   Musaceae—occasional
   Costaceae—common
Subclass ARECIDAE
1. Order Arecidae
2. Order Cyclanthales (1/0)
3. Order Pandanales (1/0)
4. Order Arales (2/0)
   Araceae—rare
Subclass LILIIDAE
1. Order Liliales (13/4)
   Liliaceae—rare
      (Incl. Amaryllidaceae)
   Iridaceae—rare
   Dioscoreaceae—rare
   Smilacaceae—rare
2. Order Orchidales (4/1)
   Orchidaceae—occasional

CLASS MAGNOLIOPSIDA (DICOTYLEDONAEAE)

Six subclasses comprise the Dicotyledons (Cronquist 1968). Extrafloral nectaries are apparently absent in the Magnolidae, rare in the Hamamelidae, occasional in the Caryophyllidae, and most abundant in the Dilleniidae, Rosidae, and Asteridae.

Subclass Magnoliidae Six orders of thirty-six families comprise this subclass. To date, there are no documented cases of extrafloral nectaries, although *Tinospora cordifolia* Miers. (Menispermaceae) has glandular trichomes on the lower leaf surface. The nature of the glandular contents and secretory products have not been determined.

Subclass Hamamelidae This small subclass of nine orders contains eighteen families, only one of which is known to have species bearing extrafloral nectaries. They are found in several species of *Ficus* on the basal portion of the lower leaf surface. They are of the Flachnektarien or Grubennektarien type.

Glandular trichomes are prevalent in the subclass, especially in the Juglandales and Myricales, however, sugars are not the dominant compounds.

Subclass Caryophyllidae   Only four orders and fourteen families make up this subclass. Extrafloral nectaries occur in three families: Cactaceae, Polygonaceae, and Plumbaginaceae. Semispherical nectaries are associated with areoles in species of *Opuntia, Cereus, Echinocactus, Coryphantha,* and *Neolloydia.* They may be in the areoles as seen in *Opuntia phaeacantha* var. *major* Engelm. or *Opuntia kleinize* DC., or positioned above the areoles as in *Echinocactus polycephalus* Engelm. & Bigelow. In addition, the floral bracts of *Cereus giganteus* secrete small amounts of nectar during the bud stage and possibly into flowering. These nectaries lack a discernible structure and are comparable to those found in the floral bracts of *Paeonia.*

Several species of *Polygonum* and *Muehlenbeckia* (Polygonaceae) have been reported as having nectaries on the lower surface of the pulvinus. These are of the Grubennektarien type. Nectaries were described on the bracteoles and on the midrib of the lower leaf surface in *Plumbago capensis.*

Subclass Dilleniidae   This large subclass contains twelve orders and seventy-one families. All orders, except Lecythidales, have taxa with extrafloral nectaries.

DILLENIALES—Of the three families in this order, only one, Paeoniaceae, has nectaries. This temperate family of herbs or suffrutescent shrubs have the gestaltlosennektarien (formless nectaries), described earlier on floral bracts enclosing developing flower buds. *Paeonia* is primarily an Old World genus with two species native to California. As far as is known, most species are capable of secreting nectar.

THEALES — Only two families, Marcgraviaceae and Dipterocarpaceae, of the thirteen in this order are known to have extrafloral nectaries. They are common in the Marcgraviaceae. Most often they are scattered on the lower leaf surface, especially the basal part. The tiny gland may be slightly sunken (Grubennektarien) or raised (Hochnektarien) and usually are parallel to and near the veins. Nectaries of this type are found in many species including *Marcgravia dubia* H.B.K., *M. rectiflora* Triana & Planch., *M. umbellata* L., and *Ruyschia crassipes* Triana & Planch. In addition, nectaries have been reported on the bracts in *Marcgravia, Norantea, Souroubea,* and *Ruyschia.*

In the Dipterocarpaceae, flattened, disc-shaped nectaries are found at the base of the lamina in *Shorea stenoptera* Burck. They are also found on the stipules and bracts. Peltate glands occur on *Anisoptera, Doona, Hopea, Monoporandra, Pentacme, Vateria,* and *Shorea.*

MALVALES — This order consists of six closely related families, four of which have extrafloral nectary-bearing species. The occasional occurrence of nectaries in the Tiliaceae is restricted largely to the leaves. They are present on the teeth along the basal portion of the leaves in several species of *Grewia* and *Triumfetta.* Nectaries are also reported on the outer surface of the sepals in *Honckenya ficifolia* Willd. In the Malvaceae, extrafloral nectaries are common. The most frequently encountered nectary in this family is in a sunken elongated cavity which is part of the midvein on the lower surface of the leaves. They occur in species of *Gossypium, Hibiscus, Urena, Cienfuegosia, Decaschistia, Ingenhouzia,* and *Kydia.* Nectaries also are found on the lower surface of the large, foliaceous floral bracts and on the outer surface of the sepals in several species of *Gossypium.*

In the Bombacaceae, nectaries can be found on the abaxial midvein of the leaf or near the base of the petiole of *Bombax* species. They also occur on the outer surface of the sepals in some species of *Bombax.*

Flattened, closely pressed nectaries occur on the basal part of the upper leaf surface on either side of the midvein in some *Sterculia* (Sterculiaceae). Taxa in other genera of this family have nectaries on the abaxial midvein, *e.g., Buettneria pilosa* Roxb. Also, nectaries are present on inflorescences near the base of the pedicels in several *Helicteres* species.

SARRACENIALES — Two of the three families in this order are known to produce nectaries. They are common in the Sarraceniaceae, a family of insectivorous plants. Tiny nectar glands are scattered on the outside of the pitcher and the ascidiform leaves in several species of *Sarracenia*, i.e., *S. flava* L., *S. minor* Wall., and *S. rubra* Walt. They have also been reported on the outer surface of the operculum in *Heliamphora*, *Darlingtonia*, and *Sarracenia*. Species of the Old World *Nepenthes* (Nepenthaceae) reportedly have nectaries on the shoot axis, petiole, lamina, tendrils, as well as their pitchers and ascidiform leaves.

VIOLALES — This is a large order of twenty-six families, six of which are known to contain extrafloral nectar-bearing species. They are rare in the Flacourtiaceae, being found along the leaf margin or lower leaf surface in a few species (*Paropsia obscura* Hoffm., *Banara mollis* Tulasne, and *Idesia polycarpa* Max.). In the Bixaceae, nectaries occur on the pedicels just below the attachment of the sepals. Stipitate glands (possibly nectaries) are present on the circinate leaves of young plants of *Triphyophyllum peltatum* (Hutch. & Dalz.) Airy Shaw of the Dioncophyllaceae.

Nectaries are common among the many vining members of the Cucurbitaceae; most are found on the lower leaf surface scattered between the veins of the basal portion of the leaf. Examples of this include *Luffa aegyptiaca* Mill., *Coccinea palmata* Cogn., *Bryonia amplexicaulis* Lam., and *Cephalandra indica* Naud. Paired petiolar nectaries are present in *Momordica cochinchinensis* Spreng. and probably other species too. Also, nectaries occur on the floral bracts in the leaf axils. Finally, they are produced on the outer surface of the sepals in both *Hedgsonia heteroclita* Kock, and female flowers of *Luffa aegyptiaca*.

The closely related Turneraceae and Passifloraceae have homologous paired nectaries at the base of the lamina or along the petiole. These nonvascularized nectaries of the Hochnektarien type are characteristic of all members of the Turneraceae and are present in many species of Passifloraceae. Flattened, disc-shaped nectaries occur in many taxa of the Passifloraceae on the lower leaf surface, mainly near veins on the basal portion of the leaf. Nectaries are also known to occur on the leaf rachis in species of *Deidamia* and on the bracts of *Passiflora incarnata* L.

SALICALES — Salicaceae, the only family in this order, contains two genera which bear nectaries, *Populus* and *Salix*. They are more common in *Populus*, occurring on the petiole or along the margin of the lamina where it joins the petiole.

CAPPARALES — Zimmerman reported nectaries occurring in the leaf axils of *Capparis cynophallophora* L. (Capparaceae). This is the only family in this order known to bear extrafloral nectaries.

ERICALES — Of the nine families in this order, only the Ericaceae has been confirmed as having extrafloral nectaries. They occur in several genera on the petiole or at the intersection of the lamina and petiole, or near the proximal edge of the lamina.

EBENALES—Nectaries are rare in the Ebenaceae, the only pertinent family of the five in this order. Flattened, usually circular nectaries are located on the lower leaf surface between the veins in *Diospyros discolor* Willd., *D. lotus* L., *D. emarginata* Hieron., and others.

Subclass Rosidae Rosidae is the largest subclass of dicotyledonous plants. It contains sixteen orders of 104 families, twenty-one of which are reported having species with extrafloral nectaries.

ROSALES — Four of the seventeen families of this order include extrafloral nectary-bearing species. Nectaries are common in the Rosaceae and Chrysobalanaceae, abundant in the Leguminosae. A gland sunken in the midvein on the lower leaf surface has been reported in *Hydrangea* (Hydrangeaceae) but further study is needed to identify the secretory product.

One or more, often paired nectaries of the Flach or Hoch type are present at the distal part of the petiole in many species of *Prunus* and some species of *Rosa*. In addition, some members of this genus have branched stipules which are nectariferous.

The Chrysobalanaceae have extrafloral nectaries homologous to those in the Rosaceae, with which it is closely related and was for many years combined. Paired nectaries are present at the base of the lamina of *Chrysobalanus icaco* L. Branched, nectary-bearing stipules and bracts are present in several species of *Hirtella* (Chrysobalanaceae), the genus that includes all species of myrmecophytes in this family. In the myrmecophilous species, the nectaries on the stipules and bracts lie in close proximity to the ant-inhabited domatia.

Nonfloral nectaries are abundant in the Leguminosae. Often they develop into large, conspicuous structures, including the largest yet described (Elias, 1974). They are present on most members of the subfamily Mimosoideae, common in the Caesalpinioideae, and less so in the Papilionoideae. Most nectaries are foliar and occur on the petiole (*Cassia* spp.), between each pair of pinnae (*Albizzia*, some *Pithecellobium*, some *Cassia*, and many others). They also are produced on the leaf midvein on *Inga adenophylla* Pittier. Most, if not all species of Australian *Acacia* have a nectary abaxially near the base of the phyllodes. Extrafloral nectaries are present on the cushion-like, compressed lateral branches on the inflorescent axis of some species of *Canavalia*, *Dioclea*, *Dolichos*, and *Phaseolus*.

MYRTALES — Extrafloral nectaries are not common in this order of thirteen families. They are reported only in the Combretaceae and Melastomataceae. Tropical vines have a greater incidence of extrafloral nectaries than do herbs, shrubs, or trees. It is not unexpected, then, to find them in some of the viny members of the tropical family Combretaceae. They occur on the lower leaf surface in *Terminalia paniculata* W. Roth & A. (lamina) and in *T. catappa* L. (veins). Paired petiolar nectaries are present in *Terminalia brasiliensis*, *Combretum argenteum* Bert., *Conocarpus procumbens* L., and *Laguncularia racemosa* Gaertn. Foliar nectaries are suspected to occur in many other species of Combretaceae.

Zimmerman reported nectaries on the outer surface of the sepals in *Memecylon floribundum* Bl. (Melastomataceae), however, their presence in the family is in need of further confirmation.

CELASTRALES — In this order of ten families only one, the Dichapetalaceae, is reported to have extrafloral nectaries. These are located near the base of the upper leaf surface in *Dichapetalum reticulatum* Engl. and *D. restita* Spruce and in *D. barterii* Engl. and *D. borneense* Merr. on the lower leaf surface.

EUPHORBIALES — Extrafloral nectaries are common in the Euphorbiaceae, the only member of the five-family order known to produce them. They occur on the lamina, e.g., *Alchorneopsis floribunda* Muell., *Macaranga hispida* (Bl.) Muell., at the base of the lamina, e.g., *Mallotus* spp. and *Endospermum peltatum* Merr., or along the leaf margin, e.g., *Baliospermum axillare* Bl. Paired or unpaired nectaries are present on the petiole of *Hura crepitans* L., *Hippomane mancinella* L., *Ricinus communis* L., *Homalanthus populifolius* Grah., and *Croton* spp. Nectaries are less common but occasionally present on the margin of floral bracts as seen in *Excoecaria bicolor* Hassk.

RHAMNALES — A small order of the three families, one, the Rhamnaceae, of which may have nectaries near the base of the leaf. They were reported occurring in *Cormonema melsoni* Rose, but this requires further substantiation.

SAPINDALES — Five of the seventeen families of this order are known to produce extrafloral nectaries. In the Anacardiaceae, paired nectaries are present on the upper petiole at or near the transition from the lamina to the petiole in *Holigarna arnottiana* Hook., *H. ferruginea* March., *H. helferi* Hook., and *H. grahamii* Kurz; nectaries also occur on stipules and bracts of these species. Within the Meliaceae, nectaries are reported on the lower leaf surface, especially near the base, the *Carapa guianensis* Abul. and *C. moluccensis* Lam. Flach nectaries are reported on the pinnate and palmate leaves in some species of *Zanthoxylum* (Rutaceae). All species of Rutaceae have glandular trichomes, especially on the vegetative parts, however, these represent another class of gland whose primary products are oils. Nectaries are common in the Simaroubaceae. They are found on the upper leaf surface on *Samadera mekongensis* Pierre, *S. indica* Gaertn., and *Quassia amara* L. Petiolar nectaries of the Flach type are present in *Cadellia pentastylis* Muell. Most if not all species of *Ailanthus* have nectaries on the auriculate margin of the leaflets.

GERANIALES — Five families comprise this small order, one of which, the Balsaminaceae, has extrafloral nectaries. These nonvascularized structures may occur in pairs at the stipular position, but more often are present along the upper edge of the petiole and at the tips of the lower teeth of the leaf margin. Foliar nectaries are more common on the Old World species of *Impatiens* than new World members.

LINALES — Nectaries are rare in two of three families of this order. They are reported at the base of the upper leaf surface in *Vantania obovata* Bth. and *V. oblongifolia* Mart. (Humiriaceae). Spiral nectaries are present on the lobes in *Vantania oblongifolia*. Foliar nectaries are also reported on *Linum macraci* Benth. and *L. oligophyllum* Willd. I have not been able to confirm these.

POLYGALALES — Extrafloral nectaries are common in the Malpighiaceae and Vochysiaceae, while occasional in Polygalaceae; these are the only three families in this seven-family order known to produce them. The Malpighiaceae contains many vining species, all probably with extrafloral nectaries. They occur on the lower leaf surface in several species of *Malpighia*, *Bunchosia nitida* Juss., *Bunchosia gracilis* Ndz., *Stigmatophyllon periplocaefolium* Juss., and *Hiptage madablota* Gaertn., *Heteropteris chrysophylla* H.B.K., and *H. nitida* H.B.K. Modified teeth along the leaf margin function as nectaries in *Tristellateia australasiae* A. Rich. and *Galphimia brasiliensis* Juss. Petiolar nectaries occur in *Sphedamnocarpus pulcherrimus* Gilg., *Banisteria argentea* Spreng., and *Heteropteris chrysophylla* H.B.K. In addition, nectaries on the bracts and on the outer surface of the calyx are well known in this family.

Disc-shaped nectaries are produced on the pulvinus of the leaf in many species of *Qualea* (Vochysiaceae). They also are present on the pedicels in *Qualea grandiflora* Mart., *Q. multiflora* Mart., and *Q. parviflora* Mart. Nectaries on the inner surface of the sepals are present in all members of Vochysiaceae.

In the Polygalaceae, they are found scattered on the lower leaf surface in many species of *Xanthophyllum*. Occasionally they are present on the petiole or along the leaf margin as modified teeth.

Subclass Asteridae   Six of the nine orders in this subclass contain species bearing extrafloral nectaries.

GENTIANALES — Three of the four families in this order rarely or only occasionally possess extrafloral nectaries. Several genera of the Asclepiadaceae, *Stephanotis*, *Hoya*, *Gymnema*, and *Sarcostemma*, have species with nectaries on the upper surface of the leaves. In the Apocynaceae, they have been reported in the leaf axils on *Allamanda neriifolia* Muell. Calcine nectaries have been reported in *Fagrae litoralis* Bl. of the Loganiaceae.

POLEMONIALES — Just one of the eight families, the Convolvulaceae, in this order is reported to produce extrafloral nectaries. They are common in *Ipomoea*, occurring on lower leaf surface, petiole, and on the pedicels just below the junction with sepals. These can be observed in *I. carnea* Jacq., *I. biloba* Forsk., *I. batatas* Poir., *I. bona-nox* L., *I. leptophylla* Torr., *I. pandurata* (L.) G. F. W. Mey., and *I. tuberosa* L., among others. (See Beckmann and Stucky, 1981; Keeler 1977, 1980.)

LAMIALES — The Verbenaceae is the only reported member of this five-family order to have extrafloral nectaries. In several species of *Clerodendron*, *Stachytarpheta mutabilis* Vahl., *Faradaya papuana* Scheff., *Gmelina asiatica* L., *Callicarpa longifolia* Lam., *Duranta plumieri* Jacq., and *Citharexylum pentandrum* Vent., nectaries are found on the lower leaf surface, especially near the veins at the base of the leaf. They also occur on the upper leaf surface in *Clerodendron blumeanum* Schauer and *Stachytarpheta mutabilis* Vahl. Petiolar nectaries are present in *Clerodendron fragrans* Willd., *C. fallax* Lindl., *Stachytarpheta mutabilis* Vahl., *Faradaya papuana* Scheff. and other taxa.

SCROPHULARIALES — Five of the twelve families of this order have taxa bearing extrafloral nectaries. In the Oleaceae, they are present in the form of glandular trichomes on the lower leaf surface. These can be seen in species of *Foresteria*, *Fraxinus*, *Olea*, *Osmanthus*, *Phillyrea*, and *Syringa*. Zimmerman reported nectaries irregularly distributed on the upper and lower leaf surfaces in *Melampyrum barbatum* Waldst., *M. pratense* L., and *Paulownia imperialis* Sieb. & Zucc. in the Scrophulariaceae. They have also been reported in this family on the bracts (*Melampyrum*), inflorescence (*Uroskinnera*), and sepals (*Paulownia*).

Nectaries are abundant in the Bignoniaceae, occurring on 90 percent or more of the taxa. They occur on the leaves, young shoots, sepals, corollas, and even developing fruits. They have been described by Elias and Gelband 1975, 1976, and Elias and Prance 1978. Tiny nonvascularized capular or disc-shaped nectaries are the commonly encountered types in this family.

In the Acanthaceae, septal nectaries occur in the genera *Barteria* and *Thunbergia*. They are also reported as occurring on the bracteoles of some species of *Barteria* and on the pedicels of *Thunbergia grandiflora* Roxb. Short, stalked nectaries are present on the pedicels in several genera of Pediaceae. These include *Sesamum*, *Pterodiscus*, *Pretrea*, and *Harpagophytum*.

DIPSACALES — The Caprifoliaceae is the only one of this five-family order known to have extrafloral nectaries. They are present on the lower leaf surface near the petiole in *Viburnum americanum* Mill. and on the petiole in *V. opulus* L. Most species of *Sambucus* have stipitate nectaries in the stipular position.

ASTERALES — Extrafloral nectaries are rare in the Compositae. Inouye (1979) reported them on the involucral bracts of *Helianthella quinquenervis*, a North American herb. Zimmerman cited several species of *Centaurea* as also having nectaries on the involucral bracts.

CLASS LILIATAE (MONOCOTYLEDONEAE)

Liliatae is divided into four subclasses (Alismatidae, Commelinidae, Arecidae, and Liliidae) comprising eighteen orders and sixty-one families. Three of the subclasses, excluding Alismatidae, are known to have species bearing extrafloral nectaries. This assemblage of 14 families consists primarily of aquatic or semiaquatic species.

In contrast to the Magnoliopsida (Dicotyledonaeae), the Liliatae has a small number of species with extrafloral nectaries. Floral nectaries, especially septal nectaries, are common and help to unify the class (Cronquist 1968). Septal nectaries are found in the Bromeliales, Zingiberales, and some Palmae. I consider the development of the septal nectary to have coincided with the evolution of pollinator-dependent flowers and to have preceded the appearance of extrafloral nectaries. The latter evolved fairly recently in some of the more specialized families and species.

Subclass Commelinidae Of the eight orders comprising this subclass, only two are known to have extrafloral nectaries: the Cyperales and the Zingiberales.

CYPERALES — Nectaries are rare in this order and restricted to the Gramineae. In *Andropogon gayanus* var. *bisquamulatus* (Hochst.) Hack. the nectaries are located in two areas: near the margin of the sheath immediately below the pseudopetiole and on the pseudo-petiolar ridges adjacent to the ligule (Bowden 1971). Nectaries were reported on the keel of the leaf sheath and on the culm nodes just below the leaf sheaths in *Eragrostis major* Host (as *E. megastachya* Link) (Mattei and Tropea 1908). Other examples of extrafloral nectaries in *Eragrostis* have been reported by Nicora (1941) in several South American species. These tiny symmetrical nectaries are nonvascularized, partially embedded in the leaf tissue, flattened or concave, and rounded or oval in shape. Often there are several nectaries arranged in a row parallel to the leaf margins. Presently, extrafloral nectaries are known to occur in *Eragrostis*, *Andropogon*, *Pennisetum*, and *Sporobolus* (Bessey 1884). Most of the species known to have these structures are native to Europe or Africa.

ZINGIBERALES — Eight specialized families comprise this order with the Marantaceae and the Costaceae known to have extrafloral nectaries. I suspect that other families in this order will be shown to have these organs. In both families, the nectaries are found on the outer surfaces of the floral bracts (*Costus* spp. and *Clinogyne dichotoma* Salisb.). The nectariferous tissue is organized not into a well-defined nectary, but rather as amorphous tissue capable of nectar secretion and sometimes identifiable by a color contrasting from that of surrounding tissues. In addition, tiny, globose, short-stalked nectaries are found at the base of the pedicels of the flowers in *Clinogyne dichotoma*. *Costus* is a pantropical genus, while *Clinogyne* is restricted to Africa. Small nectaries are embedded in the bracteoles subtending the flowers in certain species of *Calathea, Ischnosiphon*, and *Donax*. Other small nectaries are found at the junction of the wings of the leaf sheath and petiole in the Brazilian genus *Ctenanthe*.

Subclass Arecidae

ARALES — Of the two families in this order, only Araceae has been reported having extrafloral nectaries (Zimmerman 1932). Numerous tiny glands occur at the base of the cataphyll in *Philodendron cuspidatum* Koch & Bouche and *P. lindenii* Schott. This family has not been thoroughly investigated for the presence of extrafloral nectaries nor have the structure and function of those known been adequately studied.

Subclass Liliidae

LILIALES — Four of the thirteen families of Liliales are known to possess extrafloral nectaries: the Liliaceae, Smilacaceae, Iridaceae, and Dioscoreaceae. Small spurlike nectaries are present on the cataphylls of some species of *Asparagus* (i.e., *A. acutifolius*) Linn. The nectaries of the perianth of many species of *Lilium* are best treated as floral nectaries as their primary function is in the pollination process. These nectaries are located in a narrow channel formed by the unfused septa between two carpels.

In the African genus *Sansevieria*, small indistinguishable nectariferous tissue is present on the lower surface of the floral bracts. Nectar secretion commences prior to anthesis and continues into the postanthesis stage. The nectaries are especially obvious on *Sansevieria thyrsiflora* Thunb., *S. cylindrica* Boj., and *S. ehrenbergii* Schweinf. Almost identical nectaries or areas of nectariferous tissue are found in several species of *Iris* (Iridaceae). These similarities support the close relationship of the Liliaceae and Iridaceae as indicated by modern phylogenists.

In the Smilacaceae, species of *Smilax* (*S. macrophylla* Robx. and *S. bona-nox* L.) have tiny nectaries on the lower surface of the leaf blade. These are nonvascularized, flattened structures which are partially or completely enclosed in other tissue and usually rounded in shape. These are very similar to those found on the lower leaf surface in the Dioscoreaceae. Several species of *Dioscorea* have tiny nectaries of the type just described scattered near but not directly associated with prominent veins. Petiolar nectaries (adaxial surface) are also present in some *Dioscorea*). The morphological and typological similarities of the extrafloral nectaries found in the Smilacaceae and in the Dioscoreaceae imply a close relationship as opposed to inclusion of the Smilacaceae with the Liliaceae.

ORCHIDALES — This order consists of three small families plus the Orchidaceae which contains 18,000 to 20,000 species. Nectar production is prevalent in this family, functioning mostly as an attractant for pollinators. In some genera, functional nectaries are present on the outer floral parts and bracts. The nectariferous tissue is not organized into a definite structure, but is of the amorphous type described earlier. This is present on the outer surface of the sepals in *Spathoglottis plicata* Blume, *Notylia Barkeri* Lindl., and species of *Phaius* (Van der Pijl and Dodson 1966). In species of *Vanda*, *Cymbidium*, and *Grammatophyllum*, nectariferous regions are located at the base of the sepals on the outer surface. Furthermore, nectary sites are present on the outer surface of the floral bracts (often functional in bud stage) of certain species of *Cymbidium*, *Oncidium* and in *Spathoglottis plicata* Blume.

REFERENCES

Arbo, M. 1972. Estructura y ontogenia de los nectarios foliares del genero *Byttneria* (Sterculiaceae). *Darwiniana* 17:104–158.

Beckman, R. L., Jr. and J. M. Stucky. 1981. Extrafloral nectaries and plant guarding in *Ipomoea pandurata* (L.) G. F. W. Mey. (Convolvulaceae). *Amer. J. Bot.* 68:72–79.

Bentley, B. L. 1977. Extrafloral nectaries and protection by pugnacious bodyguards. *Ann. Rev. Ecol. Syst.* 8:497–427.

Bhattacharyya, B. and J. K. Maheshwari. 1971a. Studies on extrafloral nectaries of the Leguminales. I. Papilionaceae, with a discussion on the system of the Leguminales. *Proc. Indian Nat. Sc. Acad.* 37:11–30.

Bhattachyaryya, B. and J. K. Maheshwari. 1971b. Studies on extrafloral nectaries of the Leguminales. II. The genus *Cassia* Linn. (Caesalpinaceae). *Proc. Indian Nat. Sc. Acad.* 37:74–90.

Bowden, B. N. 1971. Studies on *Andropogon gayanus* Kunth. VI. The leaf nectaries of *Andropogon gayanus* var. *bisquamulatus* (Hochst.) Hack. (Gramineae). *Bot. J. Linn. Soc.* (London) 64:77–80.

Carlquist, S. 1969. Toward acceptable evolutionary interpretations of floral anatomy. *Phytomorphology* 19:332–362.

Caspary, R. 1848. *De Nectariis*. Bonn: Elverfeld.

Cronquist, A. 1968. *The Evolution and Classification of Flowering Plants*. Boston: Houghton Mifflin.

Delpino, F. 1875. Rapporti tra insetti e tra nettarii estranuziali in aloune piante. *Boll. della Soc. Entomol.* (Florenca) 7:69–90.

Dop, P. 1927. Les glandes florales externes des Bignoniacées. *Bull. Soc. Hist. Nat.* (Toulouse) 56:189–198.

Elias, T. 1972. Morphology and anatomy of foliar nectaries of *Pithecellobium macradenium* (Leguminosae). *Bot. Gaz.* 133:38–42.

Elias, T. 1980. Foliar nectaries of unusual structure in *Leonardoxa africana* (Leguminosae), an African obligate myrmecophyte. *Amer. J. Bot.* 67:423–425.

Elias, T. and H. Gelband. 1975. Nectar: Its production and function in trumpet creeper. *Science* 189:289–291.

Elias, T. and H. Gelband. 1977. Morphology, anatomy, and relationship of extrafloral nectaries and hydathodes in two species of *Impatiens* (Balsaminaceae). *Bot. Gaz.* 138:206–212.

Elias, T. and G. Prance. 1978. Nectaries on the fruit of *Crescentia* and other Bignoniaceae. *Brittonia* 30:175–181.

Elias, T., W. Rozich, and L. Newcombe. 1975. The foliar and floral nectaries of *Turnera ulmifolia* L. *Amer. J. Bot.* 62:570–576.

Frey-Wyssling, A. 1955. The phloem supply to the nectaries. *Acta Bot. Neerl.* 4:358–369.

Ianishevskii, D. E. 1941. The extrafloral nectar-glands of *Salix. Trudy Bot. Inst. Akad. Nauk. SSR*, Series 4 Eksp. *Bot.* 5:258–294. In Russian.

Inamdar, J. A. 1969. Structure and ontogeny of foliar nectaries and stomata in *Bignonia chamberlaynii* Sims. *Proc. Indian Nat. Acad. Sci.* Sect. B. 70:232–240.

Inouye, D. and O. R. Taylor, Jr. 1979. A temperate region plant-ant-seed predator system: consequences of extrafloral nectar secretion by *Helianthella quinquenervis. Ecology* 60:1–7.

Janda, C. 1931. Die extranuptialen nektarien der Malvaceen. *Oest. Bot. Zeitschr.* 86:81–130.

Keeler, K. H. 1977. The extrafloral nectaries of *Ipomoea carnea* (Convolvulaceae). *Amer. J. Bot.* 64:1182–1188.

Keeler, K. H. 1980. The extrafloral nectaries of *Ipomoea leptophylla* (Convolvulaceae). *Amer. J. Bot.* 67:216–222.

Laroche, R. 1974. Anatomic considerations of the calyx of *Adenocalymma comosum* (Lam.) A.P. DC. *Ann. Missouri Bot. Gard.* 61:53–533.

Mattei, G. and C. Tropea. 1908. Graminacee proviste di nettarii estranuziali. *Boll. R. Orto. Bot.* (Palermo) 7:113–117.

Nicora, E. 1941. Contribucion al estudio histologico de las glandulas epidermicas de algunas especies de *Eragrostis*. *Darwiniana* 5:316–321.

Parija P. and K. Samal. 1936. Extra-floral nectaries in *Tecoma capensis* Lindl. *J. Indian Bot. Soc.* 15:241–246.

Rao, L. 1926. A short note on the extrafloral nectaries in *Spathodea stipulata*. *J. Indian Bot. Soc.* 5:113–116.

Seibert, R. 1948. The use of glands in a taxonomic consideration of the family Bignoniaceae. *Ann. Missouri Bot. Gard.* 35:123–136.

Schnell, R., G. Cusset and M. Quenum. 1963. Contribution a l'édute des glandes extra-florale chez quelques groupes de plantes tropicales. *Rev. Gen. Bot.* 70:269–341.

Van der Pijl, L. and C. H. Dodson. 1966. *Orchid Flowers, Their Pollination and Evaluation*. Coral Gables, Fla: University of Miami Press.

Waddle, B. 1970, The nectaries of cotton. *Ark. Univ. Ext. Misc. Publ.* 127:25–27.

Wergin, W., C. Elmore, B. Hanny, and F. Ingber. 1975. Ultrastructure of the subglandular cells from the foliar nectaries of cotton in relation to the distribution of plasmodesmata and the symplastic transport of nectar. *Amer. J. Bot.* 62:842–849.

Zimmerman, J. 1932. Über die extrafloralen nektarien der Angiospermen. *Bot. Cent. Beih.* 49:99–196.

7

NECTARIES IN AGRICULTURE, WITH AN EMPHASIS ON THE TROPICS

BARBARA L. BENTLEY
STATE UNIVERSITY OF NEW YORK, STONY BROOK

Botanical teaching based on the temperate flora must necessarily be ill balanced and inadequate.   Van Steenis 1961.

For the elimination of risks inherent in agricultural pursuits, crop-pest control measures must be integrated according to scientific ecological principles.   National Academy of Science 1969

It has become increasingly apparent that the "advancing frontiers" of agriculture are now meshing with those of ecology (see Rabb et al. 1974). The expanding human population and the diminishing sources of petroleum and other resources currently required for food production are forcing us to seek alternative approaches to pest control, expanded or at least sustained yields, and development or redevelopment of land areas suitable for agriculture. Biologists, especially ecologists, have an important role to play in these efforts. Thus it seems appropriate to end a symposium on the biology of nectaries with a discussion on the role of nectaries in agricultural systems.

Although most work on nectaries in agriculture has been done in the temperate zone on pollination and honey production (see Beutler 1951; Free 1970; Akerberg and Crane 1966; Mittler 1962), I would like to emphasize tropical systems. It is in these areas that population growth rates are among the highest, and yet we frequently look to the tropics as the last unexploited lands of the work (see Richards 1963; Goodland and Irwin 1975; Whitmore 1975). In this discussion I will describe briefly past and current biological events in the tropics, but I would like to emphasize areas for future work in tropical agroeco-systems, and explore the role of nectaries in the pollination of crop plants and their potential in biological control systems. I certainly will make no claim to provide an exhaustive review of either agricultural research (which is enormous) or the literature on nectaries (which has been so adequately provided in this symposium). Rather, I hope the references cited here will serve to introduce nectary biologists to relevant literature on agricultural research.

SOME CHARACTERISTICS OF THE TROPICS

One of the most important characteristics of the tropics for agricultural systems is the relatively constant temperatures through-out the year, even in areas with marked seasonal differences in rainfall (Blumenstock 1958; Beckinsdale 1957; Schnell 1971; Leith 1973). At lower elevations, the temperatures range from 20°C to 30°C, and even at high elevations, the temperatures, though lower, are appropriate for many crop plants (Steenis 1968; Richards 1962b). Interestingly, the highest yields are often found at mid-elevation (1500 m) sites where daytime temperatures are high enough to allow rapid growth while the low nighttime temperatures reduce respiration rates and al-low accumulation of photosynthate (Grubb and Whitmore 1966, 1967; Grubb et al. 1963; Janzen 1973a, 1973b) which may itself be the desired product (e.g., potatoes) or may allow the plant to attain a greater biomass to support the final product (e.g., wood, fruit, etc.) (Flenley 1971).

The constancy in temperature also allows for year-round activities of insects, both beneficial species such as bees, as well as pest species (Gray 1972; Janzen 1973a, b). In a sense, the aseasonality of tempera-ture in the tropics denies these regions the "benefits" of winter during which the populations of most pest species are dramatically reduced or at least inactive (Uvarov 1964). The only comparable situation in the tropics is an extended dry season during which many insect species migrate to areas of greater water availability (e.g., riparian forests or irrigated farm land), or enter some form of diapause (Janzen and Schoener 1968). In fact, in some regions of Costa Rica, for example, a farmer may successfully grow a crop of beans only in the very early rainy season when soil moisture is sufficient for plant growth but pest insect populations have not yet recovered from dry-season decrease (Atkinson 1953; W. Hagnaur, pers. comm.). Later in the rainy season pest populations can be high enough to destroy the crop.

On the other hand, pollinators are frequently more active during periods of lower rainfall. In Guanacaste, Costa Rica, for example, many of the tree species flower only in the dry season (Frankie et al. 1974). Although the major selective forces involved in this pattern are probably related to the facts that it is probably easier for insects to fly when it is not actually raining and because pollen is more likely to be dry (Janzen 1967), nectar may play an important role not only in insect nutrition but in their water balance as well. It would be interesting to note, for example, if the sugar concentration of nectar is reduced in those species that flower at the height of the dry season. In this case, the "resource" may be water rather than sugar.

A second characteristic important to nectary biologists in tropical regions is that a very large proportion of both the natural vegetation and cultivated crops are pollinated by insects (or other animal vectors such as bats and hummingbirds) (Whitehead 1969; de NeHancourt 1977; Faegri and Van der Pijl 1971). This applies both to major plantation (cash) crops such as cacao or coffee (Dessart 1961), and also to crops such as beans, peanuts, curcubits, papaya, and other tropical fruits grown by the subsistence farmer (Geerts 1963; Tempany and Grist 1958). Since nectar is frequently the reward offered to the pollinating species, we must pay close attention not only to the characteristics of the nectar provided, but also to the biology of the pollinating species and the effects that other agricultural practices (such as the application of pesticides) may have on their abundance and distribution. (Proctor and Yeo 1975; Newsom 1967).

A related phenomenon is the high frequency of self-incompatability among tropical species. For example, a majority of the trees in Costa Rica are self-incompatible (Bawa and Opler 1975). And, while the herbacious species have not been as thoroughly studied, there are strong indications that self-incompatibility is extremely common in the tropics (Bentley 1978; Bawa and Opler 1975; Janzen, pers. comm.). The incompatibility can be due to a variety of different mechanisms from dioecy or heterostyly to timing of release of pollen vs receptivity of the stigma (de NeHancourt 1977). For example in some varieties of avacados, there are two strains, one of which releases pollen in the morning but the stigma is not receptive until the afternoon, while the other strain has the reverse pattern (receptivity in the morning and pollen release in the afternoon) (Proctor and Yeo 1975). Obviously, for most efficient fruit yield, a farmer should plant both strains in essentially equal numbers. In contrast, such dioecious crops as papaya should have a preponderance of female plants but should include enough males to provide adequate pollen. In any case, the frequency of incompatibility necessitates the maintenance of the pollinator population in whose absence pollination simply would not occur.

This brings us to the next characteristic of tropical regions which has grave implications to agricultural systems—habitat destruction. Although habitat destruction is not unique to the tropics, some of its effects may be more acutely detrimental in tropical regions than in temperate zones (Stevens 1968; Brunig 1973; Lowry 1971). As Janzen (1974) pointed out, habitat destruction can have effects ranging far beyond the actual acreage converted to pasture or bean field. For example, secondary species common to disturbed areas or as agricultural pests frequently have much broader time periods of flowering. The pollinators which normally would visit widely dispersed forest species are "lured" to roadsides, old fields, and waste areas (Free 1968). As a consequence, the forest species are not pollinated, and ultimately could go extinct. More important to the agricultural biologist is the destruction of nest sites which may accompany wholesale destruction of a forest (Free and Buttler 1959; Southwood 1971; Van Emden 1965). For a crop which is both self-incompatible and insect pollinated, the pollinators must not only be present, but abundant. As nest sites are destroyed, abundance will go down. (Way 1966; Van Emden 1965). The relationship between the proximity of a forest and yield of a crop has already been documented by Hawkins (1965) who found that yields in clover were three times greater in fields adjacent to forests than in those more than 2 km from a forest. Although this study was done in the temperate zone, we have every reason to expect that these effects exist in the tropics as well.

A final note on pollination in tropical systems has to do with "exotic" crops. Many tropical crops, notably large plantation cash crops such as cacao, rubber, or vanilla, can be grown more successfully in areas outside their region of origin. This is due primarily to the reduction of species-specific crop pests or diseases (deBach 1971). However, species-specific pollinators may also be left behind when crops are exported and grown in "foreign soil." For some crops, such as cacao (Dissart 1961), this may have very little effect as native species switch to pollinate the crop. For others, however, farmers have been forced to either abandon the scheme or, as in the case of vanilla, hire people to pollinate the flowers by hand (Proctor and Yeo 1975).

Another possibility would be to import the pollinator along with the crop (Mittler 1962). There are two major problems to be resolved prior to such an introduction: first, of course, is to determine the effects that the imported species might have on the native species, as for example the "African" bee seems to be having in South America (Taylor 1977), and second is to establish the biological requirements of the pollinator (e.g., nest site, alternate food sources, breeding behavior, etc.). Biologists have estimated that most crops need 100 to 1000 bees per acre for maximum yields (Mittler 1962). Thus any imported species must not only survive, but be at high densities, *and* stay in the vicinity of the orchard, field, or plantation.

Thus, any change in agricultural practices must include answers to the following questions:

1. Who are the pollinators? Are they species-specific? If the crop is imported, can native species be substituted?

2. What is the general biology of the pollinators? Where do they nest? Is the crop a preferred food choice? Are there alternate food sources available to maintain the population when the crop is not in flower?

3. What are the timing patterns of nectar and pollen production? Are there diurnal patterns? Are there seasonal patterns? Will the pollinators respond to these patterns?

4. What is the compatibility system of the plant? If selection for self-compatibility is a possibility, what effect would inbreeding have?

5. Is there a potential for competition for pollinators between the crop plant and other plant species in the area? Is there a potential for competition between the pollinators and other species visiting the plant?

BIOLOGICAL CONTROL

A second area where nectary biology can play a role in tropical agricultural systems is biological control of crop pests (deBach 1964; Huffaker 1971; Simmonds 1971). In this case, we are not dealing with pollinators at floral nectaries but rather ants and other hymenopterans attracted to extrafloral nectaries. Although this interaction has never been exploited commercially, the presence of predaceous, parasitic, or territorial insects on plants with extrafloral nectaries can reduce the effects of herbivores on these plants (Knipling 1970; Janzen 1967; Bentley 1976, 1977; Schemske 1978; Inouye and Taylor 1978). For example in a *Bixa orellana*, I found that seed set was significantly higher on those plants which had abundances of ants (Bentley 1976). The most effective species were *Ectotomma tuberculatum*, *E. ruidum*, and *Monasis bispinosa*, but *Crematogaster* sp, *Pseudomyrmex gracillis*, *Camponotus silciventris*, and *C. planatus* were present and possibly contributed to the protection of the plant. In a second set of experiments, common bean plants (*Phaseolus vulgaris*) were treated with droplets of "nectar" on the leaf blades (Bentley 1977). These plants had significantly higher dry weights at the end of the experimental period than did the untreated control plants.

Obviously, these data suggest that the potential of ants as biological control agents should be more thoroughly explored. However, there are a number of cautions which must be recognized. First is that the success of protection depends on the abundance of the ants on the plant (Bentley 1980; Bentley and Benson 1981). Not only does this vary from habitat to habitat, but the distribution of appropriate nest sites varies within habitat (Petal 1978; Pisarski 1978). It is sometimes possible to increase the abundances of ants on a crop by simply carrying colonies to the area (Beirne 1974; deBach and Hagen 1964; Stehr 1974). For example, leaf colonies of *Oecophylla* cultivated in southern China are carried to more northern citrus orchards for control of aphids (Rao 1972). Because *Oecophylla* cannot overwinter in the more northern locations, this process must be repeated each year.

The local abundance of ants may also be limited by alternate food sources (Petal 1978). Either the natural nectary production or the prey abundance may not be sufficient for colony growth and/or maintenance during the critical periods of crop vulnerability. However, it may be possible to increase both "nectar" and "prey" from artificial sources (Hagen and Hale 1974; Hagen et al. 1971). For example, encapsulated protein solutions have been used to supplement the diet of lacewing larvae (deBach 1964). These artificial "insect eggs" are acceptable to ants, though this system has never been used in a biocontrol system (R. Riech, pers. comm.). Ants readily come to sugar baits as well, and thus the needed sugar requirements could be met with strategically placed feeding stations. It is not appropriate, however, to simply spray a crop with sugar because increased sugar content of a leaf (either on the surface or in the tissue) can act as a feeding stimulant to herbivores (Thorsteinson 1960; Hagen et al. 1971).

Second, any use of ants as biocontrol agents must be integrated with the other pest control systems necesesary for economic exploitation of the crop (Metcalf 1974; FAO 1966; Stern et al. 1959; Ripper et al. 1951, 1949). If, for example, a crop is subject to fungal attack as well as herbivory, the fungicide cannot be toxic to the ants (or must be applied when the ants are not active or at times when the role of ants is less critical to the success of the crop). Equally, the ants cannot interfere with the other potential biocontrol agents. For example, ants might not be effective where coccinellid beetles are used to control aphids or membracids. Not only could the ants prey on the coccinellid larvae, they might also exploit the aphids or membracids for honeydew.

Third, the biocontrol system must be logistically and economically feasible (Carlson and Castle 1972; Headley 1972). In other words, the level of protection must be great enough to ensure an edible crop while at the same time be within the technical expertise of the farmer (Schumacher 1973). This is especially true for the subsistance farmer in the tropics. If a system can be developed using native ant species and simple food supplements or nest manipulations, then the probability of success would be greatly increased.

Since antiherbivore protection is greatest near the food source of the control species (Bentley 1977; Way 1963; Stradling 1978), initial research should include those crop plants such as *Gossypium* (cotton) and *Passiflora* (passion fruit) which bear extrafloral nectaries. Indeed some work has been done in this area by Cook (1905) on cotton in Guatemala. While working for the USDA, Cook learned of the Indian practice of cultivating cotton near nests of *Ectotomma tuberculatum*. This species of ant is highly predaceous and has specific behaviors for capturing mature boll weevils and other cotton pests. Cook felt that *E. tuberculatum* should be imported into the cotton field of southern Texas and used as biocontrol agents. His plan failed for two rather critical reasons: 1) he did not know the biology of the ant species well enough to insure colony survival in Texas, and 2) the leading ant researcher of the time failed to recognize the potential of ants in the protection of plants against herbivores (Wheeler 1910). Subsequent to this aborted attempt at biocontrol, DDT and other insecticides were developed and proved to provide significant control of pests on cotton. Obviously the application of such insecticides precludes the use of predaceous ant species in a pest control program (Newsom 1967; Lingren and Ridgeway 1967). In fact, researchers are now concerned about the utilization extrafloral nectar by the pest species and are currently attempting to breed for "nectarless" cotton (Benschoter and Leal 1974; Rhyne 1965, 1972; Lukefahr 1960; Lukefahr and Griffin 1956).

However, as the dangers of pesticides become increasingly apparent, more agricultural researchers are recognizing the need to develop effective biological control programs. And our current understanding of both coevolution and the biology of hymenoptera suggest that Cook's hypothesis might very well be an effective alternative.

Although protection may be greater on plants with extrafloral nectaries, the nectar source need not be on the crop plant per se. For example, control of many crop pests by parasitic hymenoptera and other predators is increased if nectary-bearing plants are grown in hedgerows adjacent to the crop (van Emden 1965; Leius 1967). The hymeoptera use the extrafloral nectaries as sugar source, but oviposit on the pest species. Since sugar is a critical component of the adult diet of the hymenopterans (Stradling 1978; Leius 1967; Thornsteinson 1960), the females will be more abundant in areas with adequate sugar resources.

Parasitic hymenoptera, in fact, can be important in reducing herbivore levels on plants bearing extrafloral nectaries as well (Knipling 1970; Leius 1967; Stehr 1974; deBach and Hagen 1964; Huffaker 1971). For example, L. Gilbert and J. Smiley (pers. comm.) have found that the nectaries on *Passiflora* are preferentially visited by microhymneopteran parasites of *Heliconius* larvae. Ants not only do not visit these nectaries, but can interfere with the parasite–host interaction by molesting the ovipositing female. However, given the high specificity of parasitic hymenoptera, they are only useful as biocontrol agents on those crop plants whose pest insects are appropriate hosts for the parasite (Clausen 1974). Ants, on the other hand, are notoriously polyphagous, and most species combine scavenging and predation with the collection "plant saps and exudates" (Petal 1978). This is true not only of temperate zone species (Ayre 1968) but tropical species as well (Lévieux 1967, 1972). This means that many ant species can be effective predators on crop pests without regard to the taxonomic status or geograhic origin of the pest (deBach 1971; Whitcomb 1974; Clausen 1974).

From the various studies cited above, it is apparent that nectaries may very well have potential in biocontrol programs utilizing ants or other species of hymenopterans. Thus, it is quite reasonable to suggest that nectary biologists can contribute significant information to the development of these programs.

Obviously the first questions to ask are related to the biology of the predator and/or parasite (Bennett 1974). We need to know its nest sites, reproductive patterns and population dynamics. And, of course, we need to know its food habits: what nectar sources can it exploit (both plant species and nectary morphology)? Is it affected by nectar composition (e.g., amino acids, "rare" sugars, etc.) How far will it travel to a nectar source? How specific are its food preferences?

Since ant abundance is positively correlated with nectar flow (Bentley 1976, 1977, 1980; Bentley and Benson 1981), it would be important to investigate the potential for increasing nectar production by the plant. Although nectar flow rates are in part a function of proximal events such as weather (Findlay et al. 1971), soil moisture (Eaton and

Ergle 1948) and isolation (Darwin 1877; Findlay and Mercer 1971b; Groner 1937), the basic pattern is probably genetically based. Instead of breeding for "nectariless" cotton, we should perhaps breed for "high yield" nectaries.

FUTURE FARMING PRACTICES IN THE TROPICS

Although farming practices in the tropics may seem far removed from the work of the nectary biologist, the application of any research must take into account the real and potential patterns of agriculture in a region. Thus I would like to conclude with the presentation of some ideas on the future of farming in the tropics and how this may influence research directions.

I feel this is especially important because agricultural research has been dominated by temperate-trained researchers who sometimes may fail to recognize the differences between tropical and temperate regions. For example, in recent decades agricultural practices in the temperate zone have led to consolidating small farms into vast single-crop "megafarms." Although plantations of cash crops do occur in the tropics (Tempany and Grist 1958), there are a number of reasons why consolidation of basic food crops should not be allowed to occur in the tropics. First is that most farms in the tropics are based on subsistence farming by an extended family group (Tothill 1940, 1948; Boserup 1965; UNESCO 1962). Consolidation would remove direct control of the land from the people to agrobusinesses. There is considerable evidence that the efficiency of the individual farmer decreases as he or she becomes further removed from decision making processes and control (Schumacher 1973).

More important for the biologists, however, is the nature of the agricultural areas in the tropics. Nowhere in the tropics are there vast areas of homogeneous soils comparable to the Great Plains of North America or the interior of the Soviet Union (UNESCO 1961; Dudal and Moorman 1964; Mohr and Schuylenborah 1972). The Amazon basin, for example, is an extremely complex mosaic of soils deposited as rivers flooded and/or changed course (Goodland and Irwin 1975). Within 1 km^2 the soils can change from essentially pure quartz sand to laterite or heavy clay. This pattern precludes the possibility of vast monoculture crops and thus, de facto, forces farmers to grow a diversity of crops on the particular soil types found in the area.

This enforced diversity may in fact be a benefit to tropical agriculture. First, there is increasing evidence that pest control, especially of insect pests, can be more effectively managed on more diverse farms (Laster 1974; Pollard 1971; Lewis 1965; deLoach 1971; Southwood 1971; Robinson et al. 1972; Pimentel 1971). This is related to the relatively high specificity to host plants among pest species but may also be related to decreased "apparency" of small plots dispersed among plots of other species. Second, the nutritional needs of the family may be more effectively met on a farm with a wider variety of crops. For example, a diet combining rice, corn and beans provides a balanced complement of amino acids required for human nutrition (Lappé 1971; FAO 1970). In addition, the family would be less dependent on the money economy and thus less vulnerable to the exigencies of global markets and the complicating factors of processing, storing, and distribution of foods from more distant areas (Schumacher 1973). This pattern does not exclude the farmers from growing cash crops but rather allows a buffer from fluctuations in the value of the cash crop.

Obviously the efficacy of this system requires relatively high yield from the acreage under cultivation. A traditional method for maintaining high yield is swidden ("slash and burn") agriculture. However, as the human population in the tropics continues to expand, this system can no longer be considered viable (UNESCO 1961, 1962). Swidden agriculture usually requires a minimum of three to seven years (Tothill 1940, 1948 in Tempany) between cultivation periods to allow the fertility of the soil to be reestablished (Nye and Greenland 1960; Spencer 1966; Coulter 1950). Even today, population pressures are forcing families to re-use land before this critical period has passed (Tempany and Grist 1958). The inevitable extension of this process could be ever-decreasing soil fertility and thus ever-decreasing yields (Anderson 1962; Webb and Williams 1972). Obviously future research must include the development of techniques to increase the probability of sustained yields from a given area of land (Thomas 1967).

Again diversification of crops can provide a partial solution. Since different crop species often have different nutritional requirements

(Coulter 1950; Harper 1977) the coexistence of two species can actually decrease interplant competition (Harper 1977). In addition, the well-known practice of crop rotation to include, for example, nitrogen-fixing species of legumes can increase soil fertility under some circumstances.

Legumes, in fact, are among the more promising of tropical crops (Tempany and Grist 1958). Not only do the nodulated forms fix nitrogen and often have protein-rich foliage and seeds, many species have extrafloral nectaries as well. Thus a crop of legumes planted among other crop species may both enrich the soil and provide sugar sources for potential predators on pest species.

Even more than sustained yields, the other two critical factors of tropical agriculture, biocontrol of pest species and crop diversification, often require labor intensive farming practices (Maxwell and Harris 1974). This would be a dramatic shift away from the current trend toward increasingly energy-intensive practices and the dependence of the petroleum industry not only for energy but for both pesticides and fertilizers. Again the potential for success of these practices is especially high in the tropics. For example, the rates of human population growth are highest in tropical regions, approaching 3 percent per year in some areas in Latin America. In addition, the extended family unit is still viable in many tropical cultures in both the Old and New World. These patterns provide both a large labor force and the organizational context in which labor-intensive small-farm agriculture can be best exploited. Thus any research in tropical agriculture should take into account the potential for labor-intensive methods.

In conclusion then, I would like to encourage all researchers to view their work in the context of world needs. Our goals should be toward integrated research between "pure" and "applied" biologists. Not only will this improve the quality of science, but may in fact improve the quality of life. Fortunately, this is especially easy for the nectary biologist.

REFERENCES

Akerberg, E. and E. Crane. 1966. Proceedings of the second international symposium on pollination. *Bee World* 47: suppl.

Anderson, J. A. R. 1962. Research on the effects of shifting cultivation in

Sarawak. *Proc. Symp. on Impact of Man on Human Tropics Vegetation.* UNESCO, Goroka.

Atkinson, B. J. 1953. The natural control of forest insects in the tropics. *Int. Congr. Entomol.* 9:220–223.

Ayre, G. L. 1968. Comparative studies on the behaviour of three species of ants (Hymenoptera; formicidae). I. Prey finding, capture, and transport. *Can. Entomol.* 100:165–172.

Bawa, K. S. and P. A. Opler. 1975. Dioecism in tropical forest trees. *Evolution* 29:167–179.

Beckinsdale, R. P. 1957. The nature of tropical rain fall. *Trop. Agric.* Trinidad. 34:76–98.

Beirne, B. P. 1974. Status of biological control procedures that involve parasite and predators. In Maxwell and Harris, eds., *Proc. . . . Biological Control of Plant Insects and Diseases* (q.v.), pp. 69–76. Jacksonville: University Press of Mississippi.

Bennett, F. D. 1974. Criteria for determination of candidate hosts and for selection of biotic agents. In Maxwell and Harris, eds., *Proc. . . . Biological Control of Plant. Insects and Diseases*, pp. 87–96.

Benschoter, C. A. and M. P. Leal. 1974. Relation of cotton plant nectar to longevity and reproduction of the cotton leaf-perforator in the laboratory. *J. Econ. Entomol.* 67:217–218.

Bentley, B. L. 1976. Plants bearing extrafloral nectaries and the associated ant community: Interhabitat differences in the reduction of herbivore damage. *Ecology* 54:815–820.

Bentley, B. L. 1977. The protective function of ants visiting the extrafloral nectaries of *Bixa orellana* L. (Bixaceae). *J. Ecol.* 65:27–38.

Bentley, B. L. 1981. Ants, extrafloral nectaries, and the vine lifeform and interaction. *Tropical Ecology* 22:127–133.

Bentley, B. L. and W. Benson. 1981. The influence of ant foraging patterns on oviposition by herbivorous insects. *Tropical Ecology*, in press.

Beutter, R. 1953. Nectar. *Bee World* 34:106–116.

Blumenstock, D. I. 1958. Distribution and characteristics of tropical climates. *Proc. Pacif. Sci. Congr.* 9:3–20.

Boserup, E. 1965. *The Condition of Agricultural Growth.* London: Aldine.

Brian, M. V. 1978. *Production Ecology of Ants and Termites.* I.B.P. 13 Cambridge: Cambridge University Press.

Brunig, E. F. 1973. Species richness and stand diversity in relation to site and succession of forests in Sarawak and Brunei (Borneo). *Amazoniana* 4:293–320.

Carlson, G. A. and E. N. Castle. 1972. Economics of pest control. In *Pest Control Strategies for the Future*, pp. 71–87. Washington, D.C.: National Academy of Science.

Clausen, C. P. 1974. A world review of parasites, predators and pathogens introduced to new habitats. U.S. Dept. Agr. *Tech. Bull. 1751.*

Cook, O. F. 1905. Habits of the Kelep or Guatamalan cotton boll weevil ant. U.S. Dept. Agr., *Bull. Bur. Entomol.* 49:1–15.

Coulter, J. K. 1950. Organic matter in Malayan soil: A preliminary study of the organic matter content in soils of divergent jungle, forest plantations, and abandoned cultivated land. *Malayan Forester* 13:189–202.

Darwin, F. 1877. On the glandular bodies of *Acacia sphaerocephala* and *Cecropia* peltata serving as food for ants, with an appendix on nectarglands of the common brake fern. Pteris aquilina. *J. Linn. Soc.* (London) *Bot.* 15:398–409.

DeBach, P., ed. 1964. *Biological Control of Insect Pests and Weeds.* New York: Reinhold.

DeBach, P. 1971. The use of imported natural enemies in insect pest management ecology. Proc. Tall Timbers Conf. on Ecol. Animal Control by Habitat Mgt. No. 3, pp. 211–233

DeBach, P. and K. S. Hagen. 1964. Manipulation of entomophagous insects. In P. DeBach, *Biological Control of Insect Pests and Weeds* (q.v.), pp. 429–458.

deLoach, C. J. 1971. The effects of habitat diversity on predators. Proc. Tall Timbers Conf. on Ecol. Animal Control by Habitat Mgt. No. 2, pp. 223–241.

deNeHancourt, D. 1977. *Incompatibility in Angiosperms.* New York: Springer.

Dessart, P. 1961. Contribution a l'etude des Ceratologonidae (Diptera). Les *Forcipomyia* pollinisateurs du cacaoyer. *Bull. Agric. Congo* 52:125–140.

Dudal, R. and F. R. Moorman. 1964. Major soils of southeast Asia. *J. Trop. Geog.* 18:54–80.

Eaton, F. M. and D. R. Ergle. 1948. Carbohydrate accumulation in the cotton plant at low moisture levels. *Plant Physiol.* 23:169–187.

Faegri, J. and L. van der Pijl. 1971. *Principles of Pollination Ecology.* London: Pergamon Press.

F.A.O. 1966. Integrated pest control. Rome.

F.A.O. 1970. Amino acid content of foods and biological data on proteins. Rome.

Findlay, N. and N. V. Mercer. 1971. Nectar production in *Abutilon*. II. Submicroscopic structure of the nectary. *Aust. J. Biol. Sci.* 24:657–664.

Findlay, N., M. L. Reed, and F. V. Mercer. 1971. Nectar production in *Abutilon*. III. Sugar secretion. *Aust. J. Biol. Sci.* 24:665–675.

Flenley, J. R. 1974. Transactions of the Third Aberdeen-Hull Symposium on Malesian ecology: Altitudinal zonation in Malesia, Univ. Hull, Dept. Geography, miscellaneous series no. 16.

Frankie, G. W., H. G. Baker, and P. A. Opler. 1974. Comparative phenological studies of trees in tropical wet and dry forests in Costa Rica. *J. Ecol.* 62:881–919.

Free, J. B. 1968. Dandelion as a competitor to fruit trees for bee visitors. *J. Appl. Ecol.* 5:161–178.

Free, J. B. 1970. *Insect Pollination of Crops*. London: Academic Press.

Free, J. B. and C. G. Buttler. 1959. *Bumblebees*. London: Collins.

Geerts, H. C. 1963. *Agricultural Involution*. Berkeley: University of California Press.

Goodland, R. J. and H. S. Irwin. 1975. *Amazon Jungle: Green Hell to Red Desert*. New York: Elsevier.

Gray, B. 1972. Economic tropical forest entomology. *Ann. Rev.* Entomol. 17:313–354.

Grover, M. G. 1937. Sugar excretion in *Impatiens sultani*. *Amer. J. Bot.* 26:464–467.

Grubb, P. J. and T. C. Whitmore. 1966. A comparison of montane and lowland forest in Ecuador. II. The climate and its effects on the distribution and physiognomy of the forests. *J. Ecol.* 54:303–333.

Grubb, P. J. and T. C. Whitmore. 1967. A comparison of montane and lowland forest in Ecuador. III. The light reaching the ground vegetation. *J. Ecol.* 55:33–57.

Grubb, P. J., J. R. Lloyd, T. D. Pennington, and T. C. Whitmore. 1963. A comparison of montane and lowland rain forest in Ecuador. I. Forest structure, physiognomy, and floristics. *J. Ecol.* 51:567–601.

Hagen, K. S. and R. Hale. 1974. Increasing natural enemies through the use of supplementary feeding and non-target prey. In Maxwell and Harris, eds., *Proc. . . . Biological Control of Plant Insects and Diseases* (q.v.), pp. 170–181.

Hagen, K. S., E. F. Sawall, and R. L. Tassan. 1971. The use of food sprays to increase effectiveness of entomophagous insects. Proc. Tall Timbers Conf. on Ecol. Animals Control by Habitat Mgt. No. 2, pp. 59–81.

Harper, J. L. 1977. *Population Biology of Plants*. London: Academic Press.

Headley, J. C. 1972. Defining the economic threshold i. In NAS, *Pest Control Strategies for the Future*. Washington, D.C.: National Academy of Science.

Huffaker, C. B. 1971. *Biological Control*. New York: Plenum Press.

Inouye, D. and O. R. Taylor. 1979. A temperate region plant-and-seed predator system: Consequences of extrafloral nectar secretion by Helianthella quinquenervis. *Ecology* 60:1–7.

Janzen, D. H. 1967. Synchronization of sexual reproduction of trees within the dry season in Central America. *Evolution* 21:620–637.

Janzen, D. H. 1973a. Sweep samples of tropical foliage insects: Description of study sites, with data on species abundance and size distributions. *Ecology* 54:664–687.

Janzen, D. H. 1973b. Sweep samples of tropical foliage insects: Effects of seasons, vegetation types, elevation, time of day, and insularity. *Ecology* 54:687–708.

Janzen, D. H. 1974. The deflowering of Central America. *Nat. Hist.* 83:48–53.

Janzen, D. H. and T. W. Schoener. 1968. Differences in insect abundance and diversity between wetter and drier sites during a tropical dry season. *Ecology* 49:96–110.

Knipling, E. F. 1970. Influence of host density on the ability of selective parasites to manage insect populations. Proc. Tall Timbers Conf. on Ecol. Animal Control by Habitat Mgt. No. 2, pp. 3–21.

Lappe, F. M. 1971. *Diet for a Small Planet*. New York: Ballantine.

Laster, M. L. 1974. Increasing natural enemy resources through crop rotation and strip cropping. In Maxwell and Harris, eds., *Proc. . . . Biological Control of Plant Insects and Diseases* (q.v.), pp. 137–149.

Leith, H. 1973. *Phenology and Seasonality Modeling*. Berlin: Springer.

Leius, K. 1967. Influence of wild flowers on parasitism of tent caterpillar and codling moth. *Can. Entomol.* 99:444–446.

Levieux, J. 1967. La place de *Camponotus aevapimensis* Mayr (Hymenoptere, formicidae) dans la chaine alimentaire l'une savane de Cote d'Ivoire. *Insects Sociaux* 14:313–322.

Levieux, J. 1972. Le role des Fourmis dans les resaux trophiques l'une savane preforestiere de cote d'Ivoire. *Ann. Univ. Abidjan*, Series E, 5:143–240.

Lewis, T. 1965. The effects of shelter on the distribution of insect pests. *Sci. Quart.* 17:74–84.

Lincoln, C. 1974. Use of economic thresholds and scouting as a basis for using parasites and predators in integrated control programs. In Maxwell and Harris, *Proc. . . . Biological Control of Plant Insects and Diseases* (q.v.), pp. 182–189.

Lindgren, P. D. and R. L. Ridgway. 1967. Toxicity of 5 insecticides on several insect predators. *J. Econ. Entomol.* 60:1639–41.

Lowry, J. B. 1971. Conserving the forest—a phytochemical view. *Malayan Nat. J.* 24:25–30.

Lukefahr, M. J. 1960. Effects of nectariless cottons on populations of three lepidopterous insects. *J. Econ. Entomol.* 53:242–244

Lukefahr, M. J. and J. A. Griffin. 1956. The effects of food on the longevity and fecundity of pink bollworm moths. *J. Econ. Entomol.* 49:876.

Maxwell, F. G. and F. A. Harris, eds. 1974. *Proceedings of the Summer Institute on Biological Control of Plant Insects and Diseases*. Jacksonville: University Press of Mississippi.

Metcalf, R. L. 1974. Selective use of pesticides in pest management. In Maxwell and Harris, eds., *Proc. . . . Biological Control of Plant Insects and Diseases* (q.v.), pp. 190–203.

Mittler, T. E. 1962. Proceedings of the first international symposium on pollination. Swedish Seed Growers Association Communication 7:1–224. Uppsala.

Mohr, E. C. J. and N. Schuylenborgh. 1972. *Tropical Soils*. 3d ed. The Hague: Mouton.

NAS. 1969. *Insect Pest Management and Control*. Washington, D.C.: National Academy of Science.

Newsom, L. B. 1967. Consequences of insecticide use on non-target organisms. *Ann. Rev. Entomol.* 12:257–86,

Nye, P. H. and D. J. Greenland. 1960. Soils under shifting cultivation. Technical Communication No. 51, Commonwealth Bur. Soil Sci. Harpenden.

Petal, J. 1978. The role of ants in ecosystems. In Brian, ed., *Production Ecology of Ants and Termites* (q.v.)., pp. 293–325.

Pimentel, D. 1971. Population control in crop systems: Monocultures and plant spatial patterns. Proc. Tall Timbers Conf. on Ecol. Animal Control by Habitat Mgt. No. 2, pp. 209–221.

Pisarski, B. 1978. Comparison of various Giomes. In M. V. Brian, ed., *Production Ecology of Ants and Termites* (q.v.), pp. 326–331.

Pollard, E. 1971. Hedges. VI. Habitat diversity and crop pests: A study of *Brevicoryne brassicae* and its syrphid predators. *J. Appl. Ecol.* 8:751–780.

Proctor, M. and P. Yeo. 1973. *The Pollination of Flowers.* London: Collins.

Rabb, R. L., R. T. Stinner, and G. A. Carlson. 1974. Ecological principles as a basis for pest management in the agroecosystem. In Maxwell and Harris, eds., *Proceedings . . . Biological Control of Plant Insects and Diseases* (q.v.), pp. 19–45.

Rao, B. P. 1972. A review of the biological control of insects and other pests in S. F. Asia and the Pacific region. *Technical Communication No. 6, Commonwealth Institute of Biological Control.*

Rhyne, C. L. 1965. Inheritance of extrafloral nectaries in cotton. *Advancing Frontiers of Plant Science* 13:121–135.

Rhyne, C. L. 1972. Linkage of indehiscent anthers and lack of leaf nectaries in *Gossypium hirsutum* L. *Empire Cotton Growers Rev.* 49:57–60.

Richards, P. W. 1962a. *The Tropical Rainforest.* Cambridge: Cambridge University Press.

Richards, P. W. 1962b. Plant life and tropical climate. In *Biometeorology,* Oxford: Pergamon Press. p 66–103.

Richards, P. W. 1963. What the tropics can contribute to ecology. *J. Ecol.* 51:231–243.

Rick, C. M. 1950. Pollination relations of *Lycoperscion esculentum* in native and foreign regions. *Evolution* 4:110–122.

Ripper, W. E., R. M. Greenslade, and L. A. Lickerish. 1949. Combined chemical and biological control of insects by means of a systemic insecticide. *Nature* 163:787–789.

Ripper, W. E., R. M. Greenslade, and G. S. Hartley. 1951. Selective insecticides and biological control. *J. Econ. Entomol.* 44:448–459.

Robinson, R. R., J. H. Young, and R. D. Morrison. 1972. Strip-cropping effects on abundance of *Heliothis*-damaged cotton squares, boll placement, total bolls, and yields in Oklahoma. *Enviroin. Entomol.* 1:140–145.

Schemske, D. W. 1978. A co-evolved triad: *Costus woodsonii* (Zingiberaceae), its dipteran seed predator, and ant mutualists. *Bull. Ecol. Soc. Amer.* 59:89.

Schnell, C. E. 1971. *Handbook for Tropical Biology in Costa Rica.* San Jose: Organization for Tropical Studies.

Schumacher, E. F. 1973. *Small Is Beautiful: Economics As If People Mattered.* New York: Harper and Row.

Simmonds, F. J. 1971. Biological control of pests. *Trop. Sci.* 12:191–201.

Soerianegara, I. 1970. Soil investigation in Mt. Hondje. F. R. W. Java, Pengun. Lemp. Pusat Penjel. Kehut. 93.

Southwood, T. R. E. 1971. Farm management in Britain and its effects on animal populations. Proc. Tall Timbers Conf. on Ecol. Animal Control by Habitat Mgt. 3:29–41.

Spencer, J. E. 1966. *Shifting Cultivation in Southeastern Asia.* San Francisco: University of California Press.

Steenis, C. F. and G. J. Van. 1968. Frost in the tropics. *Proc. Symp. Recent Adv. Trop. Ecol.* Varanasi 1:154–167.

Stehr, F. W. 1974. Release, establishment, and evaluation of parasites and predators. In Maxwell and Harris, eds., *Proceedings . . . Biological Controls of Plant Insects and Diseases* (q.v.), pp. 124–136.

Stern, V. M., R. F. Smith, R. van den Bosch, and K. S. Hagen. 1959. The integrated control concept. *Hilgardia* 29:81–101.

Stevens, W. F. 1968. The conservation of wildlife in Malaysia. Office of the Chief Game Warden, Federal Game Department, Ministry of Lands and Mines, Seremban.

Stradling, D. J. 1978. Food, and feeding habits of ants. In M. V. Brian, ed., *Production Ecology of Ants and Termites* (q.v.), pp. 81–106.

Taylor, O. R. 1977. The past and possible future spread of Africanized honeybees in the Americas, *Bee World* 58:19–30.

Tempany, H. and D. H. Grist. 1958. *An Introduction to Tropical Agriculture.* New York: Wiley.

Thomas, A. 1967. *Farming in Hot Countries.* London: Faber and Faber.

Thorsteinson, A. J. 1960. Host selection in phytophagous insects. *Ann. Rev. Entomol.* 5:193–218.

Tothill, J. D. 1940. *Agriculture in Uganda.* London: Oxford University Press.

Tothill, J. D. 1948. *Agriculture in the Sudan.* London: Oxford University Press.

UNESCO. 1961. *Proceedings of the Symposium on Tropical Soils and Vegetation.* Abidjan.

UNESCO. 1962. *Symposium on the Impact of Man on Humid Tropics Vegetation.* Goroka.

Uvarov, B. P. 1964. Problems of insect ecology in developing countries. *J. Appl. Ecol.* 1:159–168.

van Emden, H. F. 1965. The role of uncultivated land in the biology of crop pests. *Sci. Quart.* 17:121–136.

van Steenis, C. F. G. J. 1961. Axiomax and criteria of vegetatiology with special reference to the tropics. *Trop. Ecol.* 2:33–47.

Way, M. J. 1963. Mutualism between ants and honeydew-producing homoptera. *Ann. Rev. Entomol.* 8:307–344.

Way, M. J. 1966. The natural environment and integrated methods of past control. *J. Appl. Ecol.* 3(suppl.):29–32.

Webb, L. J. and W. T. Williams. 1972. Regeneration and pattern in the subtropical rainforest. *J. Ecol.* 60:675–695.

Wheeler, W. M. 1910. *Ants, Their Structure, Development, and Behavior.* New York: Columbia University Press.

Whitcomb, W. H. 1974. Natural populations of entomophagous arthropods and their effect on the argo ecosystem. In Maxwell and Harris, eds., *Proceedings . . . Biological Control of Plant Insects and Diseases* (q.v.), pp. 150–169.

Whitehead, D. R. 1969. Wind pollination in the angiosperms; evolutionary and environmental considerations. *Evolution* 23:28–35.

Whitmore, T. F. 1975. *Tropical Rainforests of the Far East.* Oxford: Clarendon Press.

8

STUDYING NECTAR?
SOME OBSERVATIONS ON THE ART

ROBERT WILLIAM CRUDEN
UNIVERSITY OF IOWA

SHARON MARIE HERMANN
UNIVERSITY OF IOWA; UNIVERSITY OF ILLINOIS

The many questions we have received with respect to techniques and the relevance of temporal patterns of nectar production to pollination studies, studies of energy budgets in birds, etc., suggest that our experiences might prove of value to those interested in the study of nectar production. Although nectar volumes and sugar concentrations are easily measured in the field, the elucidation of production patterns and the interpretation of data are less easily handled. We present ideas, some data, and many observations that are anecdotal in nature. Our desire is to aid others in pursuing their projects, illustrating things that did not work for us, as well as describing things that did work. Our discussion focuses on patterns of secretion and by and large ignores the constituents of nectar.

WHAT TO COLLECT AND MEASURE

Because the temporal pattern of nectar secretion, volume and sugar concentration of the nectar available to a pollinator at the time it initiates foraging activity (Cruden, Hermann, and Peterson 1981), and the ratio of sucrose to glucose plus fructose (Baker and Baker

1981) are strongly correlated with pollinator class, quite accurate predictions about a plant's pollinators can be made from a thorough examination of nectar and its production. A complete evaluation includes the daily pattern of secretion, the volume of nectar secreted, and an analysis of the sugars and other constituents of the nectar. How much and what types of data one collects will reflect the questions asked.

Minimally, both sugar concentration and nectar volume are required to draw any conclusions. (Note that the product of these values does not yield an accurate estimate of the sugar in a given volume of nectar: Bolten et al. 1979; Inouye et al. 1980.) For example, the flowers of *Hippobroma* (=*Isotoma*) *longiflora* (L.) Don. (Lobeliaceae), a species of low elevations in the New World tropics, are white, have long narrow corolla tubes, open in the evening and produce large quantities of nectar (range: 67–119 μl/flower). The logical conclusion is hawkmoth pollination. However, the nectar of plants grown in the greenhouse had a concentration of 10 to 11 percent, half that of typical hawkmoth-pollination plants. A pollen-ovule ratio of approximately 60:1 suggested the flowers were facultitatively autogamous (Cruden 1977) and in the greenhouse flowers set large numbers of seeds without the aid of pollinators. Using only the volume of nectar in this instance might have contributed to reaching an incorrect conclusion with respect to the pollination biology and breeding system of the plant.

Unexpected insight may accrue from the examination of nectar volume, sugar concentration, and amount of sugar per flower. Data from *Caesalpinia pulcherrima* suggest two interpretations (fig. 8.1). First, nectar secretion is continuous from initiation of secretion through mid-afternoon (curve A). Alternatively, nectar secretion ceases at approximately noon and little secretion occurs thereafter (curves B and C). The second interpretation is inconsistent with the fact that the amount of sugar per flower (fig. 8.2b) increased significantly and at approximately the same rate prior to and after 1200 h ($y = 0.32x - 2.53$, $n = 76$, $r = 0.678$, $p \ll 0.01$ and $y = 0.28x - 2.18$, $n = 49$, $r = 0.286$, $p < 0.05$, respectively). The increase in sugar concentration (fig. 8.2a) undoubtedly reflects the evaporation of water from the nectar and helps to explain the failure of nectar volumes to

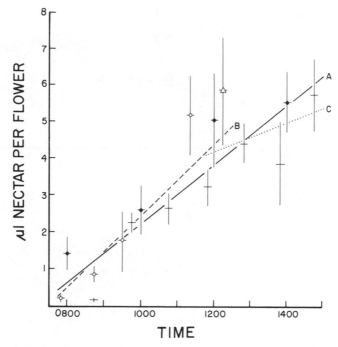

FIGURE 8.1

Nectar accumulation in hermaphroditic flowers of *Caesalpinia pulcherrima* Sw. from the Mazatlán population. $\overline{X}$'s + S.E. are given for individual data sets, i.e., all measurements made at one time. Curve A (solid line) based on all measurements y = 0.798x–5.739, n = 111, r = 0.687, P << 0.01). Curve B (broken line) based on measurements made up to and including 1200 h: y = 0.982x–7.31, n = 80, r = 0.688, P << 0.01). Curve C (dotted line) based on measurements made between 1145 and 1445: y = 0.393x–0.516, n = 46, r = 0.155, P > 0.05). Cross–14 Sep 1975; solid circle–15 Sep 1975; open circle–Sep 1974; triangle–Sep 1976.

increase significantly in the afternoon. Field observations are consistent with this interpretation. Nectar rises in a narrow tube formed by a modified petal and appears as a glistening drop at the top of the tube. As more nectar is secreted, the surface area of the exposed nectar increases, thus increasing the rate of evaporation, which should be highest in the afternoon. Two points can be made. First, examination of the amount of sugar per flower yielded the best picture of nectar

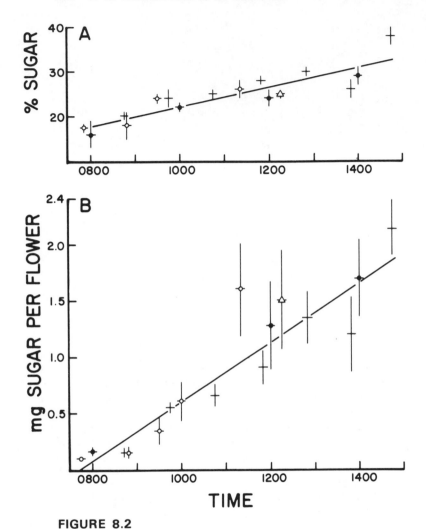

FIGURE 8.2

A. Sugar concentration of nectar from hermaphroditic flowers of *Caesal-pinia pulcherrima* in the Mazatlán population. Regression based on all measurements: y = 0.021x–0.013, n = 93, r = 0.633, P << 0.01.

B. Amount of sugar per hermaphroditic flower. Regression based on all measurements: y = 0.264x–2.039, n = 111, r = 0.714, p << 0.01.

$\overline{X}$ ± S.E. are given for individual data sets. Cross–14 Sep 1975; solid circle–15 Sep 1975; open circle–Sep 1974; triangle–Sep 1976.

secretion but obscured the fact that evaporation was occurring. Second, data on volume alone may not provide a clear picture of the pattern of nectar secretion.

QUESTIONS TO ASK

The first question to ask is whether the flowers of the plant in question open synchronously or asynchronously. We have encountered a few species whose patterns of nectar production were quite inexplicable in the context of synchronous flower opening but are easily explained in the context of asynchronous flower opening. In such species, for example, *Monarda fistulosa* L., the pattern of flower opening and the contrast between the amount of nectar available in just opened flowers relative to that in recently visited flowers may be a significant adaptive feature. Description of the pattern of secretion in flowers that open asynchronously will probably require the tagging of individual flowers. The variance in available nectar is high in such species because of the large amounts of nectar in just opened flowers.

The pattern of nectar secretion in species with synchronous flower opening can be determined with relative ease although it requires a substantial commitment of time in the field. The following questions might be considered.

1. When does secretion start?
2. Is secretion continuous or does it stop at some time?
3. Are there differences in the rate of secretion during the life of a flower?
4. Is secretion the same in all the flowers on a plant?
5. Are any of the nectar constituents resorbed at any time during the life of a flower?
6. How is nectar secretion integrated with the breeding system of the plant?

Perhaps the most critical time for the flower is that time when its pollinators become active. A reasonable reward encourages repeat visits whereas a lack of reward or minimal reward might discourage further visits by a pollinator. We have used the time of first arrival of pollinators at the flowers of a species to standardize our comparisons. This is frequently, but not always, the amount of nectar in a flower after 24 h.

To measure the amount of nectar available at the initiation of pollinator activity we have pursued three strategies.

1. Measure nectar from as many individuals as possible at the time the pollinators begin to forage. This yields the smallest amount of information but it is the quickest.
2. Sample a smaller number of individuals at frequent intervals starting well in advance of pollinator activity. Regression analysis provides rates of secretion as well as the amount available when pollinators first arrive at the flowers. Obviously the greater the number of samples and greater the sample size the more accurate the estimates.

    As with many sampling problems the size of the sample is limited by the number of investigators. Realistically, one sample requires a minimum of three to four minutes. Flowers with large amounts of nectar may require five to six minutes or longer to complete one sample.
3. Sample periodically starting shortly after the pollinators become active. If nectar secretion has ceased, there will be no significant difference between samples. Using this method we were able to detect shifts in the sugar concentration of the nectar as well as show that the amount of sugar did not change significantly after the arrival of the pollinators.

NO NECTAR

Flowers with no nectar raise the specter of a visitor that has somehow visited flowers when the likelihood of a visit is very low. For example, we have observed hawkmoths visiting flowers of hummingbird-pollinated *Penstemon kunthii* five to twenty minutes prior to the arrival of the first hummingbirds. Data derived from such flowers would give a gross underestimate of the amount of nectar normally available to hummingbirds when they become active. Yet, hawkmoths may always be present. Only careful field observations can resolve such problems.

Further, nectar production in some protandrous species differs between staminate and pistillate phases and lack of nectar may be part of the adaptive nature of the pattern. For example, the flowers of *Cuphea llavea* in the staminate phase produce large quantities of nectar whereas the pistillate phase flowers produce little or no nectar.

However, significant quantities of nectar may occur in pistillate phase flowers if they were either missed or visited early in the staminate phase by hummingbirds.

RESORPTION OF NECTAR

Most components of the pattern of nectar production are easily circumscribed; the exception is resorption. It seems quite reasonable to suppose that three processes may be at work, i.e., molecular movement with no net movement, passive movement with a net movement in one direction, and active transport against a diffusion gradient. The first process is not meaningful from an ecological or evolutionary viewpoint. The second possibility is apt to happen in old flowers that are wilting, in which case the water and/or the constituents may diffuse into the drying tissue. Only careful study can reveal if this is ecologically meaningful. Simple diffusion might also occur if physiological processes remove significant amounts of a constituent thus establishing a diffusion gradient. The third process, active resorption, especially of particular constituents, is probably the easiest to demonstrate as we (Cruden, Hermann, and Peterson, 1982) did with *Penstemon gentianoides*.

Just as in documenting other facets of nectar production it is imperative to monitor the flowers. Nectar may run out of wilted flowers, or escape from the base of the corolla if the flowers abscise without falling from the calyx.

ESTIMATES OF HOURLY AND DAILY NECTAR PRODUCTION

The species we have studied have a constant rate of secretion (see paper 3). There are reports in the literature of variable nectar production during the day. To detect such variation it is necessary to find out how much nectar is secreted each hour or fraction thereof. Pairs of flowers can be selected, the nectar in one measured, the other bagged, and the nectar measured an hour later. The difference should be the amount of nectar secreted during that hour. Many flower visitors will visit adjacent flowers sequentially, and they should hold equivalent amounts of nectar following the visit. It is our experience that most pollinators leave no nectar or a more or less constant amount of nectar in the flowers they visit if they are not disturbed. If pollinator activity is high, for example, a flower of *Caesalpinia pulcherrima* may receive a visit every three minutes, the flowers will hold miniscule amounts of nectar and the base line is essentially zero.

Because many flowers do not produce nectar continuously, estimates based on flowers bagged for twenty-four hours will be significantly less than the amounts produced by flowers that were visited several times during the day. More accurate estimates will be obtained from removing accumulated nectar from flowers periodically during the secretion cycle or determining the rate of secretion and multiplying that figure by the number of hours a flower produces nectar.

PROBABLE SOURCES OF VARIATION

Initiation and Rate of Secretion   The initiation of secretion in a population of flowers may occur over a period of an hour or more. In some instances, for example in *Gaura mutabilis*, several hours may elapse between initiation of secretion of flowers on the same scape. These flowers will, however, open at approximately the same time and probably be visited by a hawkmoth within seconds of each other. It is such lack of synchronony that accounts, in part, for the variation observed repeatedly in the samples used to calculate rates of nectar secretion (see figs. 3.2–3.4).

The second component of variation in calculating rates of secretion is the actual rate of secretion. The variation in our data on continuation and/or resumption of nectar secretion in *Erythrina breviflora*, *Penstemon kunthii*, and *Ruellia bourgaei* (see figs. 3.4 and 3.6) is due to differences in rates of secretion. In each case nectar was removed by a flower visitor and presumably the resumption or continuation of secretion was thereby synchronized. Rates of secretion are, in part, a function of flower size. Large flowers have large nectaries and produce more nectar than smaller flowers. In some instances equal sized flowers have quite different sized nectaries, for example, the nectaries of the male flowers of *Caesalpinia pulcherrima* are approximately half the size of those of hermaphroditic flowers and the rate of secretion of the former is half that of the latter (see fig. 3.3). There is also the possibility that rates of nectar secretion are, to a greater or lesser degree, independent of nectary size. In other words, equal sized nectaries could have different rates of secretion and those rates would be genetically controlled. To our knowledge this facet of nectar production has not been confirmed but it is a possible explanation for some of the patterns reported by Feinsinger (1978).

Position in an Inflorescence We have encountered significant differences between flowers on the same plant, for example between the staminate and pistillate phase of flowers of *Delphinium nelsonii* and *Cuphea llavea* and the staminate and hermaphroditic flowers of *Caesalpinia pulcherrima*. When the differences are distinct, the adaptive nature of the differences may be easily explained (see paper 3). However, intraplant differences may be less obvious and the adaptive significance less apparent. *Gaura mutabilis* provides such an example. Flowers were sampled between 1425 and 1455, approximately five hours before they were to open. The upper flowers contained three times as much nectar as the lower ($\overline{X} = 0.74 \pm$ S.E. 0.08 mg sugar, n = 15; and $\overline{X} = 0.22 \pm 0.05$ mg sugar, n = 13). Since all the flowers secreted at the same rate the difference at the time of the first pollinator visit was still significant. Our initial results in working with this species were biased because we tended to collect the upper flowers first and to collect the lower flowers in subsequent sampling periods. Thus we saw a decline in nectar production at about the time hawkmoths became active even though nectar was still accumulating in half of the flowers in the population.

Thus for one reason or another, one may encounter intra- and interplant variations, as well as variation between sampling periods (fig. 8.3). We assumed that the population had ceased nectar secretion as there was not a significant increase in the amount of sugar per flower over the period of observation. A sample of 8 flowers, one each from 8 plants would have given a more meaningful data set than 8 flowers from a single plant, if we had been interested in nectar secretion as a population characteristic and not interested in intra-plant variation.

NECTAR PRODUCTION IN CUT AND POTTED PLANTS

Inflorescences of *Leonotis nepetaefolia* R. Br. were cut and placed in water in late afternoon prior to the initiation of nectar secretion and measured the following morning (table 8.1). The sugar con-

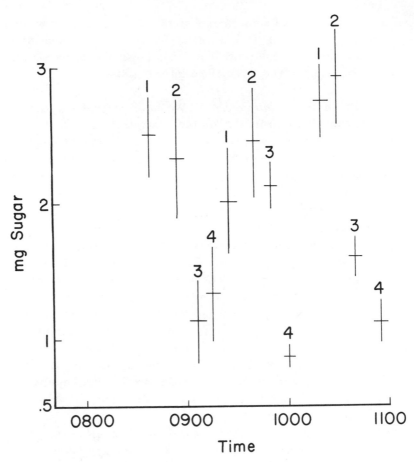

FIGURE 8.3
Milligrams of sugar in the nectar of flowers from four plants of *Ipomopsis aggregata* (Pursh.) V. Grant. $\overline{X} \pm$ S.E. are given. Each sample included eight to nine flowers.

centration of nectar from flowers from intact plants was significantly higher than that of flowers from cut inflorescences (t = 14.83, P < 0.001). The volumes of nectar in the flowers from cut inflorescences were lower but not significantly so. Thus the significant difference in the amount of sugar per flower is attributable primarily to the lower sugar concentration. A similar pattern is seen in data from *Calliandra anomala* (table 8.2).

TABLE 8.1

COMPARISON OF NECTAR IN FLOWERS FROM CUT AND INTACT INFLORESCENCES OF *LEONOTIS NEPETAEFOLIA*

	Cut Inflorescences in Water (n = 6) $\bar{X} \pm S.E.$	Intact Inflorescences (n = 12) $\bar{X} \pm S.E.$
Percent Sugar	11.1 ± 0.2	15.4 ± 0.3***
Volume (μl)	8.5 ± 0.5	9.8 ± 0.6
Sugar (mg)[a]	1.0 ± 0.1	1.6 ± 0.1**

Note: Samples from Santa María, ca 10 km N of Cuernavaca, Morelos.

[a]We assume that the amounts of amino acids were the same in these samples.

**P < 0.01   ***P < 0.001

The use of potted plants to obtain estimates of nectar production may also be subject to error. Plants from four populations of *Leonotis nepetaefolia* were grown in pots in the greenhouse. In three cases there is a small but significant drop in the sugar concentration of the nectar in the greenhouse plants compared to plants growing in the wild (table 8.3). In each instance the volume of nectar produced by plants growing in pots was significantly less than those in the wild. It follows that the amount of sugar should be less in pot grown plants. In two cases the decrease in volume is associated with a significant difference in corolla length, but in two cases there is no correlation, and in the case of

TABLE 8.2

COMPARISON OF NECTAR IN FLOWERS FROM CUT AND INTACT INFLORESCENCES OF *CALLIANDRA ANOMALA*

	Cut Inflorescences in Water[a] (N = 10) $\bar{X} \pm S.E.$	Intact Inflorescences (N = 6) $\bar{X} \pm S.E.$
Percent Sugar	8.4 ± 0.4	14.1 ± 0.8***
Volume (μl)	60.1 ± 3.3	52.2 ± 8.7
Sugar (mg)	5.2 ± 0.4	7.5 ± 0.9**

Note: Samples from ca K 69 on Ruta 95D north of Cuernavaca in Morelos.

[a]Inflorescences cut and placed in water prior to initiation of nectar secretion.

**P < 0.02   ***P < 0.001

TABLE 8.3
COMPARISON OF NECTAR PRODUCTION PER FLOWER IN NATURAL AND GREENHOUSE
POPULATIONS OF *LEONOTIS NEPETAEFOLIA*

		Barranca North of Guadalajara (1220 m) $\bar{X}$ ± S.E.	Near Tingambato, Mexico (1460 m) $\bar{X}$ ± S.E.	Cuernavaca, (ca 1600 m) $\bar{X}$ ± S.E.	East of Guadalajara (1860 m) $\bar{X}$ ± S.E.
Percent sugar	N	16.7 ± 0.4	16.3 ± 0.2	15.5 ± 0.7	16.2 ± 0.5
	G	13.9 ± 0.4***	15.4 ± 0.2**	14.3 ± 0.2NS	14.2 ± 0.4***
Volume in µl	N	13.7 ± 1.5	11.5 ± 0.7	10.7 ± 0.8	9.2 ± 0.8
	G	6.6 ± 0.6***	9.2 ± 0.5*	8.4 ± 0.5*	6.8 ± 0.5*
Mg sugar	N	2.3 ± 0.3	1.9 ± 0.1	1.7 ± 0.2	1.6 ± 0.1
	G	0.9 ± 0.1***	1.4 ± 0.1**	1.2 ± 0.1**	1.0 ± 0.1**
Flower length	N	38.4 ± 0.6	37.0 ± 0.3	38.9 ± 0.3	36.6 ± 0.6
(mm)	G	35.8 ± 0.3***	37.1 ± 0.3NS	35.9 ± 0.3***	35.8 ± 0.4NS
Sample size	N/G	9/18	13/18	9/17	15/18

Note: Measurements made in natural populations 0700–0900; in greenhouse populations, 1000–1100.
N = natural population; G = greenhouse population
*P < 0.02   **P < 0.01   ***P < 0.001

plants from Tingambato, the corollas are virtually identical in length, and still the mean nectar volume was lower in the greenhouse flowers. This suggests that environmental conditions in the greenhouse and/or pot affected nectar secretion.

A comparison of sugar concentrations and amounts of sugar from greenhouse and wild flowers of *Ruellia bourgaei* provide a strong contrast with that of *L. nepetaefolia*. The sugar concentrations are nearly the same, 22.4± 0.4 and 21.2+ 0.3, respectively, in greenhouse and wild plants, as are the amounts of sugar per flower, 60.1± 5.6 (n = 8) and 66.2± 3.5 (n = 17) milligrams, respectively. Measurements were made at approximately midnight. In this instance, plants flowering in the greenhouse produced nectar that was equivalent to that of flowers growing in the wild.

EQUIPMENT AND ITS USE

The equipment that is routinely used includes:

temperature compensated pocket refractometers
calibrated disposable micropipets   mosquito netting
washing bottle with distilled water   small ruler
facial tissues   thermometer

The pocket refractometers are used to measure sugar concentrations. Most give the percentage of sugar on a weight to weight basis, i.e., mg sugar to mg water. There are a number of refractometers on the market and they vary in price, ease of use and weight. The refractometer should be checked using a known sugar concentration or water to see that it is reading properly.

We have found that refractometers that measure concentrations from 0 to 50 percent are satisfactory for most nectars. The nectars from bee-pollinated flowers are the single exception as they frequently exceed 50 percent. A refractometer that measures in the range from 35 to 75 or 80 percent is a sound investment, a necessity if one expects to work primarily with bee nectars. In an emergency relatively accurate measurements can be made by diluting the nectar with distilled water. Such dilutions are best done, with great care, in the micropipet.

We used calibrated disposable micropipets of various volumes up to 20 microliters to collect the nectar. Micropipets of 25 and 50 microliters must be used with care as nectar may run out of them. One microliter "microcaps" proved highly successful with flowers with quite small amounts of nectar.

Calibrated hypodermic syringes were used to extract nectar from the flowers of milkweeds (Willson, Bertin, and Price 1979). Such syringes may be the way to extract nectar from quite small flowers, for example in the Compositae. Calibrated syringes are available in a wide array of sizes and may be more amenable to reuse than disposable pipets.

A third method of obtaining nectar samples from flowers that are too small for other manipulations is to place the flower(s) in water and let the nectar diffuse into the water. Kapyla (1978) used this method to obtain floral nectars for the analysis of their sugar content.

Mosquito netting is used to exclude flower visitors. Bagging of flowers or inflorescences should be done when the flowers are dry. We have bagged inflorescences in the evening and the next morning the bags were wet with dew. Wind blowing through the wet netting reduces the air temperature around the flowers (table 8.4) and undoubtedly affects the rates of nectar secretion relative to unbagged flowers. Other workers have used pollination bags, paper sacks, etc., to exclude flower visitors. We know of no experimental work, other than that given below, that addresses the question of which exclosure affects the environment of the flower the least.

We measured the temperature within three different types of bags and beneath mosquito netting (table 8.4). The temperatures were taken with YSI 42SC Tele-Thermometer. The probe was placed in direct sunlight to obtain the temperature in the sun. The results show

TABLE 8.4
TEMPERATURES ($^\circ$C) WITHIN FOUR EXCLOSURES
IN DIRECT SUNLIGHT

Time	1200 h	1700 h
Temperature in sun	26–27.5°	26–26.5°
Air speed, km/h	1.7	3
Olinkraft 2-brown bag	36–36.5°	32.5–33°
White pollination bag		
Lawson No. 217	39–40°	34.5–35°
Glassine pollination bag	41–41.5°	35.5–36°
Mosquito netting	28–30°	26.5–28°
Wet mosquito netting	21–22°	20–21°

clearly that temperatures rose least under the mosquito netting but they fluctuated more. This was a result of light air movements moving the warmed air from the exclosure. The brown bag maintained lower temperatures than the two pollination bags. In a light wind temperatures in the bags fell rapidly to ambient when a cloud came between the sun and the bag. The higher temperatures in the bags, i.e., 6 to 14°C, are sufficiently different from temperatures in the sun as to alter the rates of secretion and result in overestimates of rates of production. Although we have not measured humidity within these exclosures, logic dictates that relative humidity will be higher inside the bags than under mosquito netting.

In some instances waterproof bags might have been a great aid. For example, hawkmoth- or bat-pollinated flowers could have been bagged after sundown on misty or rainy evenings and reasonably good measurements obtained. Likewise, on overcast days bags could be used without fear of the greenhouse effect, although there would still be the potential for having biased results due to excessive humidity.

Other Equipment The ruler is used to measure the length of the nectar column in the pipet. The distilled water is used to clean the micropipets and refractometer, the latter after each measurement.

SAMPLING

In general we sampled a different set of flowers at each sampling period. This was a consequence of our inability to remove nectar effectively from long tubular corollas without damaging them. Thus, in most cases our view of nectar production is a statistical one. In those instances when we could easily and repeatedly remove nectar from a flower and do no damage to it, especially to the nectary, we did so and the results were similar to those obtained by sampling different flowers each time. Our experience suggests that repeated sampling from a flower can be done if the nectar accumulates in an area away from the nectary or if the nectary can be seen. If nectar is removed by "blind" probing of the flower, then destructive sampling may yield less questionable results.

In most instances we removed the flowers from the plant and then carefully separated the corollas from the calyx. Gentle pressure was applied to the corolla forcing the nectar into the base of the corolla from which it was removed with a micropipet. If the end of the pipet is held lower than the nectar, the nectar will run quite quickly into the pipet. If nectar does not flow easily into the pipet, the pipet is probably dirty and requires washing or should be discarded. We regularly flush the pipets with distilled water and suck or blow them dry.

We fill as many pipets as is necessary then compute the volume as follows:

$$\frac{\text{mm of nectar in the pipets}}{\text{mm equivalent of volume}} \times \frac{\text{calibrated volume}}{\text{of pipet}} = \frac{\text{volume of}}{\text{nectar}}$$

For example, if a 5 μl pipet is marked at 68 mm and two pipets are filled such that the total length of nectar in the pipets is 200 mm then the volume of nectar is:

$$\frac{200 \text{ mm}}{68 \text{ mm}} \times 5 \mu l = 14.7 \mu l.$$

CALCULATING THE AMOUNT OF SUGAR PER FLOWER

Recent papers by Bolten et al. (1979) and Inouye et al. (1980) show that calculating the amount of sugar per flower is not a simple matter of multiplying nectar concentration by the volume. There are two problems. First, because many refractometers measure sugar concentrations on a weight/weight basis, the product of the volume and sugar concentration does not yield the amount of sugar in a solution. Bolten et al. (1979) point out that the sugar concentration can be converted to g/L and this value multiplied by the volume provides an accurate measure of the amount of sugar in a volume of water. A conversion table is available in the CRC *Handbook of Chemistry and Physics*, 1978–79 (p. d-320). If you are working with raw data the use of the conversion table is both simple and rapid.

We developed a second method to enable us to work with data for which we had already computed volume times concentration. We made a dilution series of sugar solutions on a weight/weight basis, i.e., 1 mg sugar/9 ml water, 2/8, etc. The resultant volumes were divided into 10 and the quotients regressed against percent sugar concentration. The regression y = 0.0046x + 0.9946 (where x = sugar concentration) provides the correction factor. The product of the volume, the concentration, and the correction factor provides an accurate estimate of sugar per volume of water. Our estimates differ from those using the first method by 0.01 mg sugar per flower. This error probably affects the estimates less than the rounding error used in recording concentrations and estimating the volume of nectar.

The second problem involves constituents in the nectar, other than sugars, that affect the refractive index of the nectar. Inouye et al. (1980) found that amino acids contribute significantly to the refractive indices of nectar resulting in over estimates of the sugar content by 8 to 11 percent. The lipids they tested were refractively inactive. It is clear that care must be taken when drawing conclusions based on corrected refractometer readings, especially when comparing the nectars of flowers with different classes of pollinators. On the other hand there is evidence showing that the types and amounts amino acids in floral nectars are relatively constant within a species (Baker and Baker 1976) and in *Gaura mutabilis* the amounts of amino acids were the same in the nectar of young and old flowers. This suggests that valid conclusions as to the pattern of secretion can be obtained without further analysis of the nectar.

One obvious use of nectar data is to obtain estimates of the caloric value of the nectar taken by flower visitors. There are problems in addition to that of obtaining accurate estimates of the sugar content. First, amino acids are not devoid of caloric content and some, for example, proline, are used directly as an energy source by the flight muscles of various insects (see refs. in Cruden and Hermann-Parker 1979; Inouye et al. 1980). In *Caesalpinia pulcherrima*, whose nectar is rich in proline (Cruden and Hermann-Parker 1979), the effective caloric content of the nectar may be higher than that of a species whose nectar contained more sugar but less proline. Second, if a nectar contains a significant amount of lipid, which is the immediate energy source for the flight muscles of moths caloric estimates based on sugar content alone would also be underestimates.

PRESENTATION OF DATA

Perhaps the most important data to report are the volumes and sugar concentrations, i.e., the reading from the refractometer, recognizing the error inherent in the latter. It is reasonable to expect that future studies will present in tabular form, the volume and the amounts of sugar, amino acids, lipids, and other critical constituents. Such detail is necessary to understand the nutritive value of nectar to a flower visitor.

Our experience indicates that large variances are to be expected with respect to the volume of nectar and the amount of sugar per flower. The mean plus or minus the standard error should be given for each data set. The presentation of only means hides the variance, may accent samplying error, and could lead to faulty conclusions.

COLLECTING SAMPLES FOR ANALYSIS OF CONSTITUENTS

We have had great success following the recommendations of Irene Baker. We spot the nectar on filter paper, mark the diameter of the spot with four pencil marks, and indicate the volume and sugar percentage next to the spot. Be extremely careful and do not touch the area of nectar with your fingers. It will ruin the sample for amino acid analysis as fingers are amino acid rich. Keep the samples dry. We use Whatman No. 3, 9 cm diameter filter papers.

The Bakers use long strips of filter paper which they store in 35 mm film envelopes. Anyone planning to study the constituents of nectar should consult Baker and Baker (1976, and earlier papers).

ENVIRONMENTAL INFLUENCES ON NECTAR PRODUCTION

Perhaps the most appropriate observation that we can make is that the literature concerning the effect of various environmental factors on nectar secretion is terribly confused and that carefully executed and controlled studies need to be done, exploring the effects of temperature, light, and available moisture on nectar secretion. A brief summary of the impact of environmental factors on nectar secretion is given in paper 3.

REFERENCES

Baker, I. and H. G. Baker. 1976. Analyses of amino acids in flower nectars of hybrids and their parents with phylogenetic implications. *New Phytol.* 76:87–98.

Bolten, A. B., P. Feinsinger, H. G. Baker, and I. Baker. 1979. On the calculation of sugar concentration in flower nectar. *Oecologia* 41:301–304.

Cruden, R. W. and S. M. Hermann-Parker. 1979. Butterfly pollination of *Caesalpinia pulcherrima*, with observations on a psychophilous syndrome. *J. Ecology* 67:155–168.

Feinsinger, P. 1978. Ecological interactions between plants and hummingbirds in a successional tropical community. *Ecol. Monogr.* 48:269–287.

Inouye, D. W., N. A. Favre, J. A. Lanum, D. M. Levine, J. B. Meyers, M. S. Roberts, F. C. Tsao, and Y.-Y. Wang. 1980. The effects of nonsugar nectar constituents on estimates of nectar energy content. *Ecology* 61:992–996.

Käpylä, M. 1978. Amount and type of nectar sugar in some wild flowers in Finland. *Ann. Bot. Fennici* 15:85–88.

Willson, M. F., R. I. Bertin, and P. W. Price. 1979. Nectar production and flower visitors of *Asclepias verticillata*. *Amer. Midl. Nat.* 102:23–35.

INDEX TO SCIENTIFIC NAMES

SUBJECT INDEX